Early Metallurgical Sites

in Great Britain

Early Metallurgical Sites in Great Britain

BC 2000 to AD 1500

Edited by C. R. Blick

The Institute of Metals

1991

Book Number 494

Published in 1991 by The Institute of Metals
1 Carlton House Terrace, London SWIY 5DB

and
The Institute of Metals
North American Publications Center
Old Post Road, Brookfield VT 05036
U S A

British Library Cataloguing in Publication Data
Early metallurgical sites in Great Britain: B. C. 2000
to A. D. 1500.
I. Blick, C. R. II. Institute of Metals
669.09361

Library -of-Congress -Cataloging-in-Publication-Data

applied for

I S B N 0-901462-84-5

Design by Jeni Liddle in association with
PicA Publishing Services
Printed and bound by Antony Rowe Ltd, Eastbourne

Contents

Preface

In 1986 The Historical Metallurgy Society established an Archaeology Committee with the aim of providing a forum for the learned discussion of archaeometallurgical problems. Early in the life of the committee a request was received from the American Society for Metals for a list of sites of importance in the history of metallurgical development in Britain. Sadly it was realised that no such list existed, and if such a list were produced it would be a mammoth task taking some years to complete.

To reduce the effort to manageable proportions, and after discussion with American friends, it was decided to produce a list of important archaeologically authenticated sites pre-dating the time that Columbus crossed the Atlantic. David Cranstone produced an initial list and then prospective authors with detailed knowledge of the sites were approached for their views.

At this stage Charles Blick, the Conservation Officer of the Historical Metallurgy Society, was co-opted on to the committee and given the task of commissioning the articles on the individual sites which are presented in this volume. With characteristic vigour, Charles Blick produced guidance notes for the authors, edited the manuscripts and obtained much of the illustrative material and in all these tasks he was ably assisted by Peter Crew and David Cranstone, whilst Chris Salter prepared the material in a form suitable for publication. Professor R.F. Tylecote was a founder member of the Archaeology Committee and took an active part in the initial discussions leading to the preparation of this volume. It is sad that he is no longer with us to see the task completed. We are grateful to these gentlemen and the individual authors for their efforts.

I hope that the volume which has been produced will be of interest to all those who wish to see the visible remains of the work of the early metallurgists who did so much to advance the materials base of our civilisation.

Professor Jack Nutting
Chairman of The Archaeology Committee

Acknowledgements

The Society wishes to thank most sincerely the following authors and establishments whose contributions have made this book possible, and acknowledges with gratitude the permission granted to quote from published material.

Royal Archaeological Institute, Dr G. Astill, C.R. Blick & J. Blick, Dr B.C. Burnham, Cambridge University Library, Dr H.F. Cleere, D. Cranstone, P. Crew, Mrs J. Day, N. Dibben, G. Downs Rose, R. Ellam, D. Gale, Dr T.A.P. Greeves, W.S. Harvey, J.S. Hodgkinson, S.J.S. Hughes, N. Johnson, Professor G.D.B. Jones, C.A. Lewis, Dr J.G. McDonnell, Miss J. Massey, The National Trust, D. de L. Nicholls, Great Orme Exploration Society, Peak District Mines Historical Society, J. Pickin, Professor N. J. G. Pounds, Professor P. Rahtz, C.J. Salter, R.T. Schadla-Hall, M. Davies-Shiel, Dr E. A. Slater, Somerset County Council Planning Department, I.J. Standing, Dr P. Swinbank, S. Timberlake and Dr L.M. Willies.

Unless otherwise credited, the drawings are the work of Miss J. Massey, based on the illustrations produced by the individual authors, to all of whom the Society is much indebted and acknowledges permission to reproduce. The abbreviation 'NGR' has been used for 'National Grid Reference' throughout the text.

NAME/LOCATION OF SITE	TYPE OF SITE	DATING
1. NEWBRIDGE, East Sussex	*Site of the earliest British blast furnace*	AD 1496
2. TAW RIVER, Devon	*Tin blowing mill*	15th Cent.
3. PUZZLE WOOD, Forest of Dean, Gloucestershire	*Iron ore mine*	2-4th Cent.
4. DOLAUCOTHI, Pumpsaint, Dyfed	*Gold mine*	Roman
5. BORDESLEY ABBEY, Redditch, Worcestershire	*Monastic metal working area*	12th Cent.
6. CWMYSTWYTH, Copa Hill, Dyfed	*Copper ore workings*	Early-Middle Bronze Age
7. BRYN Y CASTELL, Ffestiniog, Gwynedd	*Iron smelting site*	Iron Age
8. ALDERLEY EDGE, Cheshire	*Copper ore workings*	Bronze Age
9. BRANTWOOD, Cumbria	*Bloomery*	14th Cent.
10. BROWNGILL, Cumbria	*Lead/silver ore workings*	11-12th Cent.
11. PEAK DISTRICT, Derbyshire	*Lead ore smelting sites*	Roman
12. LOSTWITHIEL, Cornwall	*Duchy Palace and Stannary*	13th Cent.
13. CHARTERHOUSE-ON-MENDIP, Somerset	*Lead ore workings*	Roman
14. LEADHILLS, Lanarkshire and WANLOCKHEAD, Dumfriesshire	*Gold workings, lead later*	15th Cent.
15. GREAT ORME HEAD, Llandudno, Gwynedd	*Copper ore working*	Bronze Age

(Distribution map shown on p.10)

9

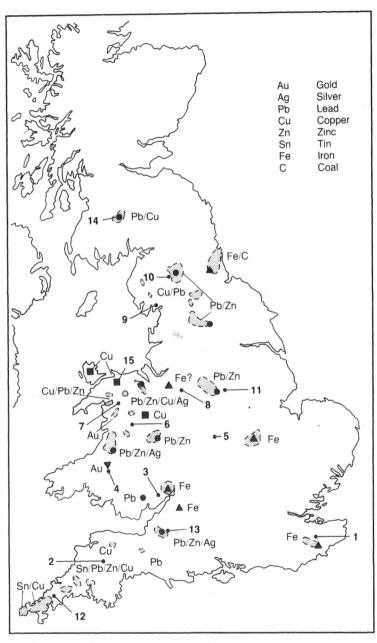

Au   Gold
Ag   Silver
Pb   Lead
Cu   Copper
Zn   Zinc
Sn   Tin
Fe   Iron
C   Coal

14 ● Pb/Cu

Fe/C

10 ●
Cu/Pb
9
Pb/Zn

Cu 15
Cu/Pb/Zn
Fe?   Pb/Zn
Pb/Zn/Cu/Ag   8   11
7
Cu
Au   6
Pb/Zn   ● 5   Fe

Pb/Zn/Ag

Au
4   3   Fe
Pb ●
Fe

13   Fe
Pb/Zn/Ag   Fe   1

Cu
2
Sn/Pb/Zn/Cu   Pb
Sn/Cu
12

Distribution map of mining sites in Great Britain.
Courtesy Professor G. D. Barri Jones

10

Introduction to the
Metallurgical Sites to 1500 AD

Archaeology, the study of man's past, utilizes techniques from a wide number of other disciplines. In 1836 one of the earliest archaeologists, C.J. Thomsen, proposed the Three Age System relying on changes in technology, particularly metals technology, to distinguish cultural changes in prehistory (i.e. the Stone Age, Bronze Age and the Iron Age). Even today with the use of independent dating methods, such as carbon-14 dating, this chronology has remained a fundamental concept in archaeology. The importance of metals in archaeological studies has been maintained and, recently, with the advent of sophisticated analytical techniques their role has increased in importance. Associated with the development of these studies interest has grown in mining and metal-working sites.

The British Isles are particularly well provided with metalliferous ores, both ferrous and non-ferrous. Although Britain was one of the last areas of Europe to adopt the use of metals, in certain periods throughout the last 5000 years it has been one of the most important centres of production in Europe.

The first metals to be used in Britain were native gold and copper. Their introduction is associated with the Beaker Culture, and the Wessex Culture which are especially rich. At the same time the Megalithic structures of Stonehenge and Avebury were being developed. This short-lived period is termed the Chalcolithic, i.e. the Copper Age.

In the last few years positive evidence of Bronze Age mining of copper (and therefore smelting) has been discovered; the earliest has been dated to 1800 BC (see Cwmystwyth). There has as yet been no equivalent firm evidence for the extraction of tin, the metal mixed with copper to produce the alloy bronze. This alloy was used throughout the Bronze Age (2000-700 BC) for manufacturing all types of artefacts, although in the Late Bronze Age lead was added in increasing amounts, partly for metallurgical purposes (it made the alloy easier to cast), but also perhaps for economic reasons.

The transition from bronze to iron occurred in Britain in the period between 800 and 550 BC. The reasons for this transition are unknown; possibly due to an interruption in the supply of non-ferrous metals. Other changes are also recorded in the archaeological record during this period, notably a change to a wetter climate. Iron ores are widespread in Britain, and there is evidence for the smelting of iron in most counties in England at some period or other. It would seem that these were ideal conditions for the rapid spread of iron-smelting and working throughout Britain. However, there was not a rapid growth in the use of iron during the Early Iron Age. The archaeological record would seem to indicate that it took several centuries for the amount of metal in circulation to recover to that of the Late Bronze Age. At the beginning of the Iron Age iron was only used for high-status objects, but it slowly began to be used for a wider range of artefacts, especially weapons. The quantities of iron recovered from excavations of Iron Age sites is low compared with later periods, although the slags derived from blacksmithing are found in small quantities on most settlement sites. Therefore, iron was either not used for everyday artefacts or was carefully recycled, the former being more probable as Iron Age nails are rare. The non-ferrous alloys were still in use, and often iron and copper alloy were used for different parts of an object, for example, the horse trappings from Gussage All Saints (Dorset).

The latter part of the Iron Age was a period of change, with strong Belgic influence entering Southern Britain from the continent, culminating in the Roman invasions of Caesar (55 and 54 BC) and Claudius (AD 43). About this time the first literary references are made to Britain, and the Roman writer Strabo noted that the island was famous for exporting slaves, hunting dogs and iron (Geography iv 199). It is therefore tempting to argue that the smelting and smithing of iron was a major industry.

All the metals, gold, silver, copper alloys (bronzes and brasses), lead and iron, were used on a large scale in the Roman period, and all could be mined and manufactured in Britain. There is extensive evidence in the archaeological record for the working of all metals, but only iron-smelting sites have been excavated, although the presence of lead pigs is indirect evidence of lead mining and smelting. The evidence indicates extensive central organization of mining and smelting, with major industrial centres being established, for example, the Weald (Sussex and Kent) and the Forest of Dean (Gloucestershire) for iron, and Derbyshire for lead. The production of artefacts was also on a large scale, and this period may be considered the first to use 'mass production'.

The end of the Roman period (c. AD 400) is marked by invasions of Angles, Saxons and Jutes. Some of these were violent, but others were probably peaceful immigrations. With these incursions come changes in metal-working technology and artefact form and decoration. Although this period used to be referred to as the Dark Ages, archaeology has shown that it was not a period returning to 'barbarism', although there were significant changes to society. Some of the finest pieces of metalwork found in Britain date from this period. There is, at present, no coherent picture of metals technology since there were many different traditions present at this time, from the Picts in the north to the Saxons in the south. The one trend that is apparent is a move away from the large centres of iron production of the Roman period to an industry based on local supplies to meet local needs. The quality of the metalwork suggests skilled craftsmen providing artefacts to order rather than the 'mass production' of the Roman period.

Production of metals slowly expanded during the medieval period, · and the centres of iron production were re-established. It was also a period of innovation, both in smelting and metalworking, with a wider range of artefacts made in metal. The church had great influence on the production and use of metals. For example, the newly established monasteries (especially the Cistercians) were often endowed with mining and smelting rights. The monasteries were carefully documented and their cartularies contain much metal-working information. Our knowledge of the medieval period derives from the massive amount of documentary sources available for study. Although most are legal documents there is a great deal of information to be extracted concerning mining and smelting rights, infringement of laws, and the paying of rents in metal, etc. Iron was used on a large scale for tools and weapons, and lead for plumbing and roofing, while bronze continued to be used for many items.

During the medieval period the extraction and production of metals played an important part in the development of the landscape, most notably in the way that woodland was carefully controlled and harvested to provide the vast amounts of charcoal required for the metal-working industries. During the 15th and 16th centuries developments were made that laid the foundation of the Industrial Revolution 200 years later, notably the replacement of the iron bloomery process by the charcoal blast furnace. Despite the abundant documentary references, the site evidence for mining and smelting of non-ferrous metals has as yet been little studied.

Prehistoric Period

The evidence that enables us to reconstruct the past history of metals technology comes from many varied sources. The majority derives from the study of slags and artefacts, not the extraction sites themselves. However, we know from several sites in Wales that during the Bronze Age copper was won from deep mines and this has been confirmed by carbon-14 dating. The surface spoil heaps visible at CWMYSTWYTH have also been dated to the Bronze Age.

ALDERLEY EDGE is another early copper mining site, although much of the evidence has been disturbed by nineteenth-century workings. No prehistoric workings have been identified at Caerloggas, where tin slag was recovered associated with pottery dated to 1600 BC. The site has now been destroyed by the modern china clay industry but the slag proves that tin was being smelted in Cornwall in the Bronze Age.

The Iron Age started about 700 BC and it seems that ore deposits were soon exploited in many parts of the country. We have the surface mines at PUZZLE WOOD in the Forest of Dean and we know that bog iron ore was used at the hillfort site at BRYN Y CASTELL, where the smith worked in a snail-shaped smithy.

Roman Period

The mineral resources of Britain were perhaps one of the incentives for the Roman invasion (though all too few of their workings can now be identified).

The MENDIP lead ores at CHARTERHOUSE were smelted within a few years of their arrival, and were exploited in later times as shown by evidence of medieval and later workings. Unlike the ores in Derbyshire the Mendip ores contained appreciable amounts of silver, which were extracted by cupellation. Pigs of lead from Somerset and Derbyshire were exported to the continent as well as being used in Roman Britain for plumbing and roofing, and for pewter tableware in conjunction with Cornish tin. Gold was mined at DOLAUCOTHI, but there is no evidence that any of the British alluvial gold deposits, such as LEADHILLS and WANLOCKHEAD in southern Scotland, were worked by the Romans.

Medieval Period

Evidence of medieval lead mining can be seen in the northern Pennines at BROWNGILL in Cumbria. Here the ore was argentiferous and its mining was fostered by the Crown. The mining was conducted at first by means of 'bell-pits', the upcast from which is represented by collars (rings) round the back-filled pits. It was normal to fill the previous pit with the spoil from the next. The bell-pits went down to the vein, or 'flats', where they widened out to follow the ore, a technique used in Neolithic flint mines such as Grime's Graves in Norfolk. This mining method required no timbering but was limited in lateral penetration, hence the need for a succession of bell-pits.

The establishment of monasteries in the Middle Ages led to both a growth in extraction of minerals and the dissemination of metal-working techniques. The monks selected sites in river valleys where they could make the maximum use of water. Water power was an important innovation (for example, in the fulling of cloth), and this could have led directly to the use of the tilt hammer in the forging of large iron blooms. The bloom size had increased by an order of magnitude from the 10kg of the Roman to the 100kg of the monastic period.

BORDESLEY ABBEY, where only the slightest indications can now be seen on the surface, was one such site. Those visiting the site should inspect the later, adjacent needle mill which is unique in having all its machinery still in working order.

The bloomery process was excellent for the making of iron in one direct step. But by the 15th century the indirect process was introduced whereby cast iron is produced, and then converted into malleable or wrought iron by a second process. (This technology was first used in China 1500 years before it was adopted in Western Europe.)

The earliest blast furnace sites are in the south east of the country, in Kent and Sussex, and are from the 15th century. Ashdown Forest was one of the few areas in this region which had sufficient fuel and water. Charcoal continued to be used as fuel for smelting iron ore (and other ores) since the sulphur which adversely affects metal quality could not be removed from coal.

NEWBRIDGE was the first recorded site to have a blast furnace. The foundations of slightly later furnaces can still be seen at Pippingford and Panningridge where the name 'Steel Forge' is an indication of some steel-making method. Most of the cast iron produced in times of war went into the production of guns, but in times of peace it was converted into wrought iron in nearby forges.

Since wrought iron was what people generally wanted it was still found economic to make it by the direct process. Bloomeries were still being established in parts of Britain where pressure on charcoal was not so great; the remains of bloomeries at BRANTWOOD are a typical example. In this process a good deal of iron was lost in the slag, which is mainly the black silicate, fayalite. Heaps of this slag can be seen all over the country; later much of it was re-smelted in blast furnaces where lime was added to displace the iron.

The evidence for non-ferrous metalworking is more varied. Unlike iron and lead, tin ores occur in low concentrations and have to be dressed by stamping and washing. On Dartmoor and in Cornwall we find the remains of many of the old mills in which this was done. The one on the TAW RIVER near Okehampton has the remains of both the dressing and smelting plant. Smelting was done with the aid of turf and charcoal in small shaft furnaces, not more than 30cm wide internally, blown by air from water-driven bellows. The same waterwheel could drive both the bellows and the stamps. The evidence for these processes are a rectangular building now open to the sky with a wheelpit, mortar stones in which the stamps worked, and mould stones. There are many of these on Dartmoor, where activity largely ceased in the 16th century, and some in Cornwall.

The main tin mining areas were soon brought under state control as their products were of national importance. This involved giving certain rights to the miners and improving quality control of the product. In the case of the latter, the ingots had to be taken to the Stannary towns where their quality could be assessed by cutting a piece off the corner of each ingot. After this was done, and quality assured, they were re-melted and stamped. Officially no tin could bypass this operation. One Stannary town was LOSTWITHIEL where the remains of the palace of the Earls, and later the Dukes of Cornwall, and where coining was done, still survive. In due course the Duchy of Cornwall became the property of the reigning monarch and its revenues were assigned to the heir to the throne.

J. G. McDonnell

SITE No. 1
Newbridge

Newbridge

1 Name of site:

Newbridge Furnace and Forge, Newbridge, Hartfield, East Sussex. Scheduled Ancient Monument.

2 Metal:

Iron.

3 General history of area:

The site lies on Ashdown Forest, a sterile area of upland reminiscent of the Pennine Moors. Formerly Lancaster Great Park and therefore a royal preserve, the area has been occupied since mesolithic times and on its surface are the marks of man's long use of the area. At Garden Hill, a fortified enclosure to the south west of Newbridge, there is evidence of neolithic, Iron Age and Roman occupation and of ironworking in the latter two periods. A Roman road crosses the Forest and its path is marked where it passes beside the B2026 between Camp Hill and the Five Hundred Acre Wood. To the south, towards Nutley, excavation revealed a small Middle Saxon ironworking site.

Ashdown Forest was a hunting ground, outside the common law, in the Middle Ages. Its boundary, the park pale, can still be traced and its form, a bank with a ditch on the forest side, betrays its purpose as a deer-leap fence, preventing the game from escaping from the forest but allowing them to enter. Gates or 'hatches' allowed people to enter and exit. In some of the valleys, lodges were built for the rangers who preserved the game for their royal visitors. Place names record their sites. Also to be seen are the long 'pillow' mounds of the rabbit warrens. Again their names continue in use.

The establishment of Newbridge Furnace restored ironworking to the Forest. Later ironworks include a Steel Forge built c. 1509, and a furnace at Stumblewood Common, on the Forest's western edge. Attempts by the Crown to sell off the Forest were the cause of bitter resentment and a decree of 1693 secured 6400 acres of grazing rights for the commoners. The rest was enclosed, including the land on which Pippingford Furnace was built on the old Steel Forge site in 1696.

Land improvements to the Forest started in the early 18th century but gathered pace in the 19th century when new farms were created at Pippingford and Crowborough Warrens. Ashdown Forest's wild and rugged character attracted the British Army who began using it as a training area during the First World War. A considerable area to the south west of Newbridge is now owned or leased by the Ministry of Defence. In a lighter vein, Ashdown Forest has become immortalized in the classic children's stories of Winnie the Pooh, by A.A. Milne.

4 Remains and dating:

Bay Length - 180m; Height 2-3m. Breached by road and Newbridge Mill Leat (part of Millbrook); partly removed west of road. West end forms a semi-circle, part of which was probably designed to protect the working area from spillway flooding. Two gaps in the semi-circular portion may indicate inlets to wheelpits.

Water System. Pond dry. Present restored spillway probably on original site. Two dry hollows within the semi-circular part of the bay, with dry ditches to main stream, may indicate wheelpits and tailraces.

Working Area. The semi-circular portion of the bay contains forge cinder and bloomery-type tap slag. North of destroyed length of bay, next to road, is a scatter of glassy slag and charcoal. Large quantities of glassy slag are known to have been removed from small field to north.

Dating. Earliest reference 1496; Latest reference 1603.

5 Location:

NGR TQ456325. Along footpath to west of minor road between Coleman's Hatch and Duddleswell, just south of Newbridge.

6 Accessibility:

Open to the public throughout the year.

7 Ownership:

East Sussex County Council. Administered by the Board of Conservators of Ashdown Forest, Ashdown Forest Centre, Wych Cross, Forest Row, East Sussex.

8 Permission required to visit:

None required. Public open space.

9 Sketch plan of site:

See page 20.

10 History of working at the site:

Newbridge is the earliest documented blast furnace site in England. It was set up for the Crown by Henry Fyner, a Southwark goldsmith, by an order of December 1496, to produce iron for Henry VII's artillery on his Scottish campaign. Iron was being produced early in 1497 and the works were leased, at £20 a year, to Peter Roberts, alias Graunt Pierre, a Frenchman. Roberts defaulted in his payment of rent and was imprisoned. By the end of 1498, Pauncelett Symart, another Frenchman, held the lease.

Newbridge Ironworks: area of Blast Furnace hearth.
Copyright © Jeremy S. Hodgkinson

Newbridge Ironworks.
Courtesy of Leicester University Press, 1985

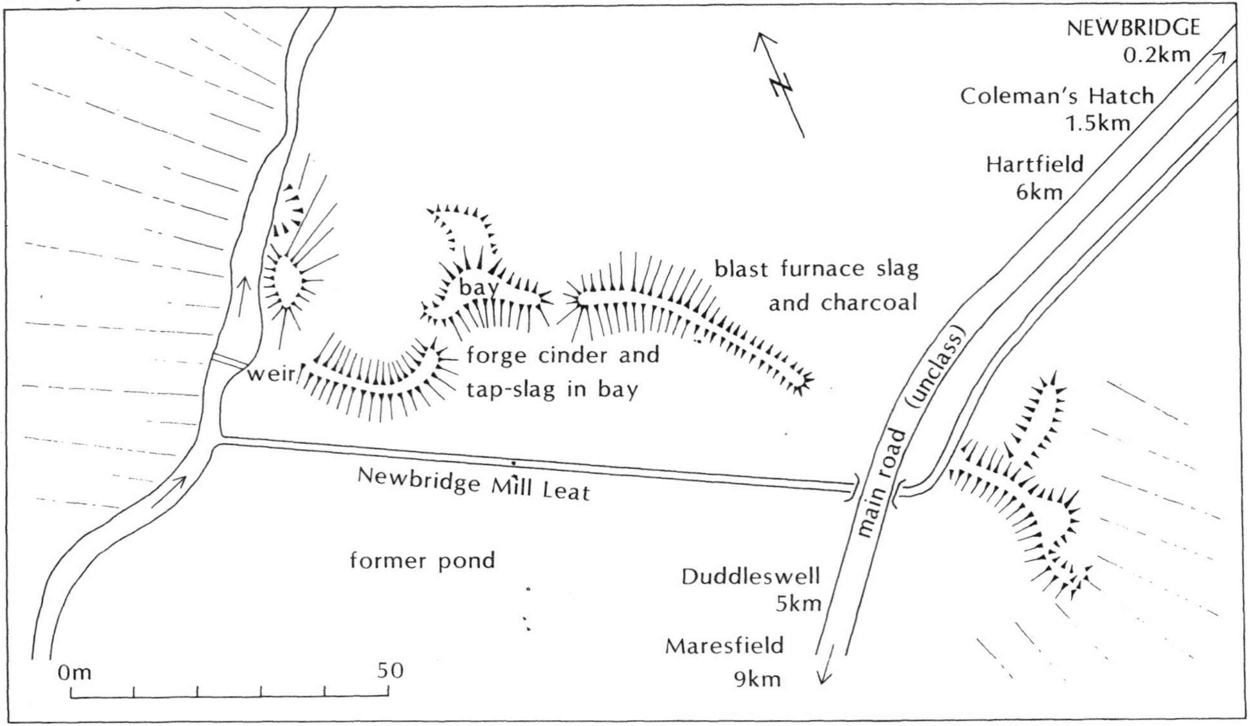

NEWBRIDGE 0.2km
Coleman's Hatch 1.5km
Hartfield 6km

main road (unclass)

Duddleswell 5km
Maresfield 9km

blast furnace slag
and charcoal

forge cinder and
tap-slag in bay

bay

Newbridge Mill Leat

former pond

weir

0m   50

Products of the furnace, listed in the accounts, included bolts, bolsters, strake bars for axles, cross bars and nails; all components for gun carriages. Also mentioned are two-part cannon. For converting the cast iron produced at the furnace into wrought iron, it is clear that a forge with water-powered hammers existed, although later accounts suggest that it was not immediately adjacent. It is possible that the forge provided the location for Simon Ballard who made gunstones (or cannon balls) from iron cast at Newbridge, and sent them to the Tower of London.

An inventory of Newbridge ironworks was drawn up in 1509 when a commission was appointed to inquire into their poor state. In 1512, Symart gave up his lease and a new one was granted to Humphrey Walker, the king's founder. The work appears to have been in decline again in 1519, and may have been out of use. They were re-let in 1525 to Sir Thomas Boleyn, the father of the future queen. Simon Forneres, the king's gunstone maker, sub-leased the site from Boleyn by 1534, but five years later it was in the hands of William Nysell, and casting about 160 tons of iron annually.

In the 1574 lists, Henry Bowyer had a royal furnace and forge on Ashdown Forest; in one version of the list this is identified as a double furnace at Newbridge. The last reference is in 1603.

11 References:

Straker, E. (1931) Wealden Iron, Bell, London.
Schubert, H.R. (1952), 'The first English blast furnace', J. Iron & Steel Inst., 170, 108-110.
Cleere, H.F. and Crossley, D.W. (1985) The Iron Industry of the Weald, Leicester University Press, Leicester.

12 Adjacent sites of interest:

BEAUPORT PARK, Battle/Westfield, East Sussex (TQ786146). Largest Romano-British iron site in the Weald. Dated by finds to AD 200-400. Slagheaps, substantial bath house. Access by public footpath (further access: contact Dr G. Brodribb, Stubbles, Ewhurst Green, Robertsbridge, Sussex).

CHITCOMBE, Brede, East Sussex (TQ 813211). Large Romano-British site. Dated by finds to AD 200-300. Slag heaps. Access by public footpath from Chitcombe Farm (further access with permission of Mr and Mrs Slinn at Chitcombe Farm).

BARNDOWN, Ticehurst, East Sussex (TQ663293). Large Romano-British site. Dated by finds to AD 200-300. Tiles connect site with Classis Britannica (as at Beauport Park). Slag heap (and abundant pottery). Access by public footpath (further access by arrangement with Dr H.F. Cleere, 'Acres Rise', Lower Platts, Ticehurst, Wadhurst, Sussex).

CROWHURST PARK, Crowhurst, East Sussex (TQ 769136). Extensive Iron Age and Romano-British site. Dated by finds. Probable iron ore quarry in The Dell on north west of the site. Access by public footpath (further access by permission of Mr Butler, Park Farm, Breadsell, Crowhurst, Battle).

ASHDOWN FOREST CENTRE (TQ 432324).

SHEFFIELD PARK GARDENS (TQ 412240) NT.

STANDEN (TQ 390356) NT.

Information from Jeremy. S. Hodgkinson, The Wealden Iron Research Group, 7 Kiln Road, Crawley Down, Crawley, Sussex RH10 4JY.

SITE No. 2
Taw River

Taw River

1 Name of site:

Taw River Tin Blowing Mill, Belstone, Dartmoor, Devon.

2 Metal:

Tin.

3 General history of area:

The tin industry of Dartmoor, which was flourishing in the early 16th century, smelted its ore in blowing mills, i.e. structures containing small shaft furnaces operated with a forced draught. The sites of scores of tin mills survive on moorland Dartmoor, mostly in a very ruined state (Greeves, 1981 a,b, 1985). Several are marked on the Ordnance Survey maps as 'Blowing House'.

Most of the mills, many of which combine the functions of stamping (crushing of ore under water-powered hammers) and blowing (smelting), are likely to date to the 16th or 17th century. It is much more difficult to identify positively those sites that were in operation before AD 1500, although smelting techniques are unlikely to have altered significantly, except perhaps in scale, since the 13th century.

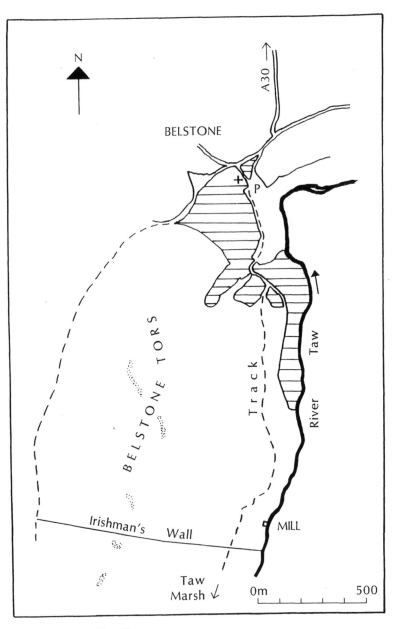

Belstone Tors and River Taw.
Copyright © Dr T.A.P. Greeves

Taw River mill has been chosen as a representative medieval tin smelting site because it appears to be documented in the 1530s and there is inferential evidence for its likely use in the 1480s, if not earlier. Furthermore, structural remains survive, the site is easily approached from the village of Belstone, and the moorland setting is impressive.

4 Remains and dating:

The Mill (see page 26). Situated at 350m (1150ft) above sea level, the mill structure is within a few metres of the left (west) bank of the River Taw. Typically, it is set on a relatively level area of ground (approx. 45m x 15m) defined by a steep but short scarp on its west and south sides. This scarp may itself have been formed by the activities of earlier tinners removing alluvial deposits. The mill structure is tucked under the scarp at the south end of the complex, enabling water to be brought to it from a higher level to power a waterwheel.

Like all mill sites of this age it is very ruined, but a careful look will reveal several interesting features.

The mill building consists of a drystone-walled rectangular structure approximately 9m x 4m internally, and with its long axis roughly N-S. An entrance gap is visible in the north-west corner where a shallow groove in a stone is the slot for a wooden door jamb. The walls still stand to a maximum height of 1.2m above the jumbled floor surface.

In the NE corner of the structure is a recessed area measuring approx. 2.3m x 0.5m (marked w/pit on map on page 26). This is the wheelpit for a small waterwheel that would have powered stamping machinery for crushing the tin ore and also probably bellows for the furnace. Only a few Dartmoor tin mills are known to have internal wheelpits such as this; most have an external one, often outside a gable end. Water for the wheel, which was probably overshot, would have been brought along a leat embankment which can be traced southward above and behind the pit for about 6m.

The headweir of the leat must have been at approx. SX 61909158, about 450m south of the mill itself. The leat is not visible at this point but it can be traced for about 300m along its course to a point about 40m from the mill. Here it has a wide and deep channel which may have been used by the tinners as a storage reservoir. There is no trace of the leat between this point and the mill, and as the slope to the river is very steep here it is possible that the tinners carried the leat on a wooden launder on this last stretch.

From the NE corner of the mill a channel leads to the river. It is 30m in length and about 1m wide, and is filled with loose stone. It clearly formed the tailrace from the wheelpit and must have connected with the latter by means of an underground culvert. At the point where the channel reaches the river there is a stone lying within the river with a hollow visible on one surface. The hollow measures 0.28m x 0.18m and has a depth of 0.08m. It

Taw River Tin Blowing Mill.
Copyright © Dr T.A.P. Greeves

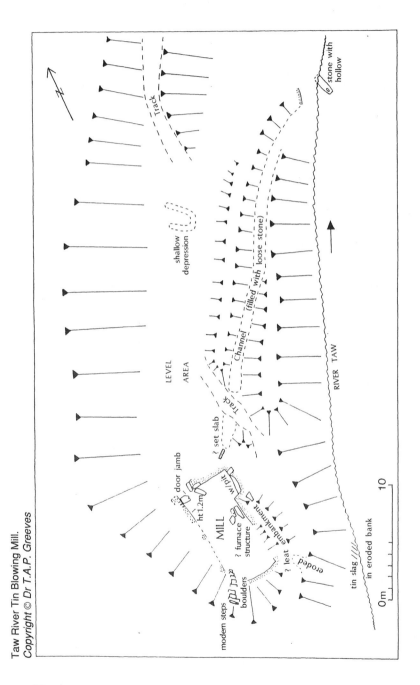

The head of Taw Marsh, about 1.5 miles south of Taw River Mill, looking south towards Steeperton Tor in c. 1900.
Copyright © Devon Library Services P & D 3563

is most likely to be a natural waterworn stone but it is possibly a mortarstone on which tin ore was crushed, the hollow having been formed by the pounding action of the stamps operated by the waterwheel.

The fact that tin-smelting took place at this mill has been established by the finding of small quantities of tin slag in the eroded river bank ESE of the mill. Also of great interest is a group of stones within the mill itself which are likely to be the remains of the furnace structure. Tin-smelting furnaces of this date seem to have been free-standing, and constructed within a space about 0.5m x 0.5m defined on two opposing sides by substantial blocks of stonework. Tin ore, crushed to a fine sand, was placed in the furnace with charcoal (peat or wood) and fired with a forced draught for about 12 hours to produce an ingot of tin metal weighing between 100-200 pounds (larger ingots were produced from the mid-16th century onwards).

At the SW corner of the mill a flight of rough steps leads down into it. These have nothing to do with the original structure, being superficially placed among jumbled boulders, and must be a modern addition to the site.

On the north side of the mill structure, the level area is likely to have been the site of settling pits, known as buddles, where crushed ore was concentrated. A curious shallow depression may have some connection with these. At the north end of the complex a track can be seen leading down to the site. This may have been the original access route.

5 Location:

NGR SX 62059197. The Ordnance Survey map Outdoor Leisure 28 (Dartmoor) 1:25000 scale is the most useful.

6 Accessibility:

Warm waterproof clothing and stout shoes or boots are recommended at almost all times of year! N.B.- no stones should be moved or other disturbance made of the area.

The site is on the northern edge of Dartmoor and can be reached from the village of Belstone which lies a short distance south of the A30, east of Okehampton.

Park your car on the east side of the village centre (marked P on the map on page 24) where open ground falls steeply to the River Taw. Follow the tarmac road southwards for 500m before taking a good but rough track out onto the open moor. After about1 km the site of the mill can be found on the east side of the track, on the bank of the River Taw.

The track continues southward to Taw Marsh where it peters out. A circular route involving rough moorland walking can be taken by following the 'Irishman's Wall' westward up and over a tor-studded ridge and then following a good track on the west side of Belstone Tors back to Belstone village.

The site, the suggested access routes to and from it, and Taw Marsh lie outside the Okehampton Range Military Training Area, so access is possible at all times of the year. However, if more extensive exploration of northern Dartmoor is planned, firing times must be checked (Tel: 0626 832093). Red flags are flown on prominent hills when firing is in progress.

7 Ownership:

Open to the public.

8 Permission required to visit:

None required.

9 Sketch plan of site:

See page 26.

10 History of working at the site:

In both 1535 and 1538-39 Thomas Takfeld was the tenant of 'a mill called Kakking Mill, and a mill called Blowing Mill (with) two acres of land lying in the Forest in the west part of the water of Tawe', for which he paid 3d rent (Moore and Birkett, 1890, p.45). Seventy years later, in 1608, a Nicholas Tuckfield paid 3d rent for a mill within the Manor of Lydford. The type of mill is unspecified but is very likely to be the same site (Duchy of Cornwall, London/Dartmore Proceedings 1203-1735, fol.29).

The valley bottom of the moorland River Taw was extensively worked for alluvial tin, especially in the huge basin of Taw Marsh which lies only 1km south of the mill site. The earliest reference to tinworking here dates to 1487 when 'Tawmersshe' was listed as one of four tinworks which had previously been sold to William Furse by Robert Parker, chaplain, who had inherited them from his parents John and Elizabeth Parker (Devon Record Office/DD 32193b). In about 1470, John Rapchyn, a tinner of Belstone, had conveyed a half share in his tinworks on Dartmoor to the same John Parker, and Taw Marsh is very likely to have been among these (Westcountry Studies Library, Exeter, Burnett-Morris index). Throughout much of the sixteenth century profits were being made from tinworks on Taw Marsh by the churchwardens of Chagford (Osborne, 1979).

These references to tinworking lend inferential support to the date of the Taw River Mill as it is only a short distance downstream from Taw Marsh and is the only known blowing mill on the moorland River Taw.

The site was first correctly identified as a mill by Eric Hemery (1983, p.847), although his description is not entirely accurate. He also quotes a lease of the mill to Thomas Takfield in 1521, but the present writer has not yet been able to confirm this source. On 25 June 1985 a plan of the site was made by Tom Greeves and Rosemary Robinson and this, combined with the discovery of tin slag, confirmed its status as a blowing mill.

11 References:

Greeves, T.A.P. (1981a) The Devon Tin Industry 1450-1750: An Archaeological and Historical Survey. Unpublished PhD Thesis, University of Exeter.
Greeves, T.A.P. (1981b) 'The archaeological potential of the Devon tin industry', in Crossley, D.W. (ed.), Medieval Industry, Council for British Archaeology Research Report 40, 85-95, London.
Greeves, T.A.P. (1985) 'The Dartmoor Tin Industry- some aspects of its field remains', Devon Archaeology, 3, 31-40.
Hemery, E. (1983) High Dartmoor- Land and People. Robert Hale, London.
Moore, S.A. and Birkett, P. (1890) A Short History of the Rights of Common upon the Forest of Dartmoor and the Commons of Devon, Dartmoor Preservation Association Publications 1, Plymouth.
Osborne F.M. (1979) The Church Wardens' Accounts of St. Michael's Church, Chagford 1480-1600, Trowbridge. Privately published.

12 Adjacent sites of interest:

FINCH FOUNDRY, STICKLEPATH, SX 640940 1.5 miles from Belstone. A nineteenth century water-powered edgetool mill.

MUSEUM of DARTMOOR LIFE, 3 West Street, Okehampton (SX 587952), 3 miles from Belstone. Nineteenth-century Dartmoor mining tools, etc.

LYDFORD (SX 510847), 12 miles from Belstone. Fascinating ancient settlement which was minting coins in the 10th century. Administrative centre for the Royal Forest of Dartmoor. The medieval castle contains the Stannary prison for miscreant Devon tinners. Fine church, good pub, good teashop. Lydford Gorge (National Trust) is an added attraction.

WHEAL BETSY MINE, Devon (SX 510815). Lead, silver and zinc in 19th century. Dramatic enginehouse.

WHEAL FRIENDSHIP MINE, Mary Tavy, Devon (SX 505795). Copper, lead, iron and arsenic mine; remains of condensing plants and a 91m (300ft) flue.

TIN BLOWING MILLS at Merrivale (SX 55277624) and (SX 55277535), about 23 miles from Belstone. Leave car in car park about 350m east of Merrivale Bridge. Cross road and enter rough moorland through gate. The very ruined mills are beside the river Walkham two-thirds of a mile and one mile upstream. Probably sixteenth-century or early seventeenth-century granite mouldstones visible at each, while the lower mill has remains of a furnace structure.

NINETEENTH CENTURY TIN SMELTING HOUSE at Eylesbarrow, Sheepstor (SX 59186765) about 30 miles from Belstone. Leave car at end of tarmac road, one and half miles east of Sheepstor, at SX 579673. Cross stream and follow rough track past plantation on your right and on to open moorland for about two-thirds of a mile. Site lies short distance south of track. Very ruined smelting house in operation 1822-1831 producing 276 tons of metallic tin. Both a reverberatory and a blast furnace were used.

TINWORKINGS at Birchtor and Vitifer near Postbridge (SX 680810), about 18 miles from Belstone via Moretonhampstead. Most impressive tinworkings on moorland Dartmoor can be seen to south of Moretonhampstead-Postbridge road near Warren House Inn. Linear gullies can be seen running approximately East-West on the hillsides - these are medieval openworks. Extensive mines were in operation in the 19th and early 20th centuries.

EXPLORATION OF THE OLD SURFACE WORKINGS MUST BE MADE WITH GREAT CARE. THEY ARE DANGEROUS !

Information from Dr T.A.P. Greeves, 17 Godstone Road, St Margarets, Twickenham, Middlesex, TW1 1JY.

SITE No. 3
Puzzle Wood

Puzzle Wood

1 Name of site:

Puzzle Wood (Scowle Holes), Near Coleford, Forest of Dean, Gloucestershire.

2 Metal:

Iron ore.

3 General history of area:

The Forest of Dean had an iron industry from the Iron Age to the late 1800s. The limonite ores occur within a narrow band of the Carboniferous Limestone known as the Crease Limestone.

Iron ore mining and smelting were probably continous throughout the two and half millenia. There were major peaks in production in the Romano-British times, during the middle to late Middle Ages, with the charcoal blast furnace from 1600 to 1800 and the coke blast furnace from 1800 to around 1870.

The dip of the Crease Limestone is around 10 degrees at Puzzle Wood giving rise to the broad outcrop which contained numerous ore bodies. Direct dating is wanting but the likely age of extraction is from Romano-British to Medieval. Romano-British settlement sites (AD 100-400) are known 0.7km north with Romano-British smelting at Coleford and 0.6km southwest.

Ancient mine workings, or "Scowles".

4 Remains and dating:

The Scowle Holes of Puzzle Wood are weathered opencasts from early iron ore mining. They are of profound natural beauty and carry a vegetation of ancient woodland which is of high botanic interest.

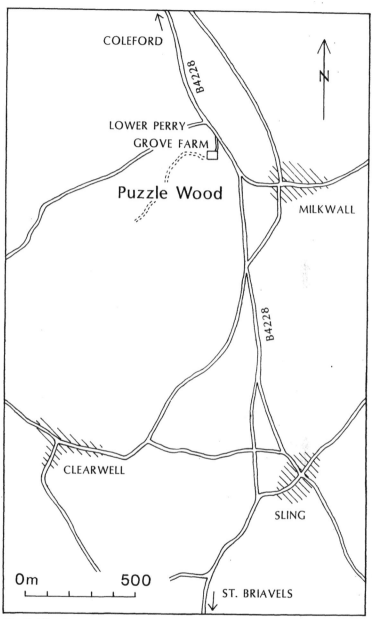

COLEFORD

B4228

N

LOWER PERRY
GROVE FARM

Puzzle Wood

MILKWALL

B4228

CLEARWELL

SLING

0m 500

ST. BRIAVELS

Puzzle Wood.
Courtesy of The Forestry Commission

5 Location:

NGR SO 580092 (Road Entrance): Address: Lower Perrygrove Farm, Coleford, Glos. The farm lies on the B4228 road about 1.5km south of Coleford town centre.

6 Accessibility:

Parking for cars and coaches is available at Lower Perrygrove Farm and an admission fee is charged. From the car park the site is approached on foot. Rustic paths laid out as a maze in the 19th century provide good foot access.

Open to the public (admission fee) Spring to Autumn each year.

7 Ownership:

Mr and Mrs Prosser, Lower Perrygrove Farm, as in section 5.

8 Permission required to visit:

Open to the public much of the year, see sections 6 and 7.

9 Sketch plan of site:

See page 33.

10 History of working at the site:

See sections 3 and 4.

11 References:

There are no specific papers. A good resumé of the early iron industry in the Forest of Dean will be found in Bulletin of the Historical Metallurgy Group,(1968), 2(1).

12 Adjacent sites of interest:

WHITECLIFF FURNACE, Coleford (David Mushet 1808-1810)NGR SO 568102.

DARKHILL FURNACE, Milkwall (David Mushet 1818-1847)NGR SO 590087.
FOREST STEELWORKS (Robert Mushet 1840-1860)NGR SO 588090.

FOREST HOUSE, Coleford (The Mushet Family Home)NGR SO 575104.

CLEARWELL CAVES (underground iron ore mines open to the public)NGR SO 578083.

LYDNEY PARK IRON ORE MINE (conclusive dating to the Roman Period) with very limited access via trap door.

Information from Ian J. Standing, Rock House, Bowens Hill, Coleford, Gloucester, GL16 8DH.

SITE No. 4
Dolaucothi

1 Name of site:

Dolaucothi Gold Mines, near Pumpsaint, Llanwrda, Dyfed.

2 Metal:

Gold primarily copper, lead, zinc associations.

3 General history of area:

At the gold mine of Dolaucothi is a wealth of surface and underground evidence spanning periods of exploitation from the Roman occupation up to the present day. They are situated about 1km SE of Pumpsaint village, where the main workings can be traced for over 1km running NE to SW along the SE side of the Cothi valley. The Romans mined here from soon after the conquest for over 200 years, developing a sophisticated water system for hushing in the opencast workings. Two access tunnels have been dubbed Roman because of their smooth finish which preceded the advent of explosives. Subsequently, there could have been small-scale operations from the 12th century to the 17th century. In 1844 gold was rediscovered on the site. Since then several companies have made attempts to exploit the mineral, the last attempt being terminated in 1943.

Mitchell Open Cast Dolaucothi Gold Mines.
Copyright © National Trust and Kathy de Witt

4 Remains and dating:

Pre-nineteenth century features can be considered under three headings:-

(1) Mine Workings. Mainly located on the northern slopes of Allt Cwmhenog, Allt Ogofau and the intervening saddle. Extremely difficult to date on the surface evidence alone.

(a) Ogofau Workings (Area 2-4 and 8, shown opposite). These occur at the central section of the lode zone and seem to have formed the principal focus of mining activity at all periods. Best approached from the main car park. The main Ogofau Pit (4) is the largest of the known opencasts, though its original dimensions have been much obscured by nineteenth and twentieth century dumping. At present, it is about 24m deep from the upper lip and measures 150m east-west by 100m north-south, but recent drilling suggests a rubble infill up to 12.5m deep in places. Currently houses the visitor centre and several reconstructed mine buildings, including a metal headframe above the original 1930s shaft. Roman miners apparently pursued the mineralized shales and richer veins underground at this point, because a fragment of a Roman drainage wheel was found during tunnelling in 1935 in an area of deep stoping.

Associated with this main opencast are three smaller pits known as Davies cutting (8), the so-called 'Roman Pit' (3) and the Mitchell Pit (2). The former might well have resulted from the collapse of an underground stope, but the other two provide useful illustrations of the opencast tradition at Dolaucothi. Roman Pit preserves much of its original outline, despite being partially infilled by a late nineteenth century dump, while Mitchell Pit is even more informative, reflecting the way in which the miners had exploited two main quartz veins. Immediately above the latter's back wall runs the line of the Cothi Leat and one of its tanks.

Various adit levels of the 19th and 20th century were driven into the walls of the Ogofau pit, or into the hillside above. These are known as Mill Adit, Middle Adit, Long Adit and Mitchell Adit, most of which are accessible as part of the National Trust's summer visitor programme. Long Adit is connected to Mitchell Adit by a vertical shaft, while stopes from the latter penetrate through to the surface in Mitchell Pit.

(b) Allt Cwmhenog Workings (Area 1). A cluster of old workings now badly obscured by forestry plantations. Comprise two coalescing opencasts, the lower one cutting into the base of the upper and containing a deep slot and two adits. Around the rim of the upper pit are the traces of several probable hushing gullies served by one or more tanks associated with the Annell Leat. Workings also appear to pre-date the line of Cothi Leat.

(c) Niagara Workings (Area 5). Another large opencast best approached from the Pumpsaint-Caeo road. Up its slopes run the visible remains of the tramway linking the 1930s shaft to the mill plant.

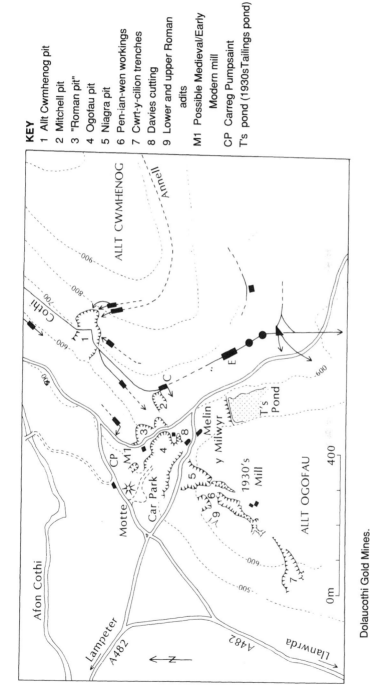

KEY

1 Allt Cwmhenog pit
2 Mitchell pit
3 "Roman pit"
4 Ogofau pit
5 Niagra pit
6 Pen-ian-wen workings
7 Cwrt-y-cilion trenches
8 Davies cutting
9 Lower and upper Roman
   adits
M1 Possible Medieval/Early
   Modern mill
CP Carreg Pumpsaint
T's pond (1930s Tailings pond)

Dolaucothi Gold Mines.
Copyright © Dr B.C. Burnham

39

(d) Pen-lan-wen Workings and Roman Adits (Areas 6 and 9). An extensive area of workings, focused on a single deep trench and an almost circular opencast with a steep rock-cut back-wall (6). The trench may have been artificially deepened by collapse into underground stopes. The northern slopes of Allt Ogofau are covered by numerous old dumps and shallow surface explorations. The ore-bodies exploited by the surface trench were also intersected at depth by a series of adits, whose entrances lie on the northern flank of Allt Ogofau. Those of Upper and Lower Roman Adits are now accessible as part of the National Trust's self-guided trail (9). First described in 1767 and surveyed in 1844, they both run for about 50m and were hand-driven as the pick and chisel marks on their roofs and walls demonstrate. Lower Roman is coffin-shaped, while Upper Roman has a square profile. Both presumably served a series of underground stopes much like the visible example at the end of Upper Roman. Usually assigned on scant evidence to the Roman period.

(e) Cwrt-y-Cilion Trenches (Area 7). A complex network of shallow trenches including two deeper pits and two short trial adits.

(2) Water Supply. The extensive provision and use of water is one of Dolaucothi's major features, usually assigned on various grounds to the Roman period. Surface remains of the two principal leats and their associated tanks and reservoirs are visible at several points, both in the mine workings and along their carefully engineered courses.

(a) Cothi Leat. Taps River Cothi 11km upstream at Hell's Pool, where rock-cut sections are visible in the gorge (SN 718466). Best seen on hillside above Llwyn-y-Ceiliog Farm (SN 686429). Traceable in the mine area swinging around Allt Cwmhenog Workings en route for Mitchell Pit, above which lay Tank C. This measured 24m by 6m. It was rock-cut into the hillside at the back, with an 8m wide outer bank made up of laminated clay and shale with a waterproof clay lining. A sluice gate in one corner originally served a rock-cut overflow channel, but was later remodelled to supply a set of three stepped washing tables. From Tank C, the leat ran on to Tank E, the largest on the site, measuring 42m x 10m and capable of holding an estimated quarter of a million gallons. It, too, was rock-cut at the back, with a 17m wide outer bank of shale, rubble and clay to resist the water pressure.

(b) Annell Leat. Runs into the workings at a higher level than the Cothi leat, but much damaged in the mine area except for the hushing channels served by one or more tanks in the forestry plantation above the Allt Cwmhenog Workings. Two deep gullies run downhill and then subdivide into at least eight lesser channels before plunging into the opencast. It was originally thought to tap the River Annell 7km upstream, but so much of its original course is now lost that certainty is impossible. Identified stretches are visible with permission on private land at NGRs SN 700426

and SN 675405. Recent survey has also identified what may be an extension of the leat a further 3km beyond the Annell to tap the river Gwenlas.

(3) Related Processing Areas. These remain poorly known, with the exception of the possible water-driven mill located on the sloping ground above the Carreg Pumpsaint stone. Identifiable features include a header tank and several feeder leats supplying water to a possible wheel pit. Likely to belong after about AD 1200 for reason stated in section 10.

(a) *The 1930s Mill Plant.* This lay on the hillside near Pen-lan-wen Farm, south of the Ogofau and Niagara opencasts. Surviving remains include the concrete foundations of several buildings, including the winder house and tippler station, the power house and a mill plant itself. The resulting tailings then passed to a tailings pond in the valley below, extensive traces of which still survive.

5 Location:

NGR SN 663403. Just off the A482 road from Llanwrda to Lampeter about 1km SE of Pumpsaint.

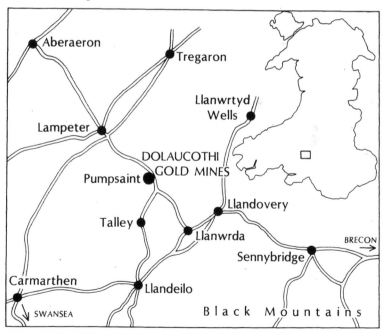

Dolaucothi Gold Mines. General plan of area.
Courtesy of The National Trust

6 Accessibility:

National Trust Visitor centre open April to October every day, with underground tours June to September. (See section 8).

Visible stretches of the Annell leat are on private land and permission is required from the relevant farmers.

7 Ownership:

The mine workings are in the ownership of the National Trust, The King's Head, Bridge Street, Llandeilo, Dyfed.

8 Permission required to visit:

None required for access to the site, but see National Trust Information Centre on site.

9 Sketch plan of site:

See page 39.

10 History of working at the site:

Three main phases:-

(1) Pre-nineteenth century mining activity. This is difficult to date on purely archaeological grounds, but the single most important phase in the development of the mines is almost certainly the Roman period. Traces of earlier (Iron Age?) exploitation are possibly represented by several circular huts associated with a quartz exposure on the southern slopes of Allt Ogofau. Roman activity apparently involved a basic progression from prospecting through opencast working to subsequent underground stoping, all associated with large-scale water provision serving various functions. This sequence remains difficult to date precisely, but its initial development is likely to coincide with the establishment of the nearby fort at Pumpsaint about AD 70. Mining more than likely outlasted the fort's evacuation around AD 160, being leased perhaps to civilian concerns. Post-Roman exploitation has been hard to trace on the ground, though the existence of a possible motte suggests one line of enquiry. More interesting are the possible remains of a water-driven mill complex near the Carreg Pumpsaint, which itself resembles a mortar stone from a trip-hammer mill. On technological grounds, such a complex ought to belong in the period c. AD 1200-1700, to be linked perhaps with the well known sixteenth and seventeenth century mining interest in West Wales.

(2) Late nineteenth and early twentieth century mining operations. Rediscovery of gold at the site in 1844 led to limited exploitation in the 1850s and 1870s. In 1888, the South Wales Gold Mining Company was formed and work was concentrated on the so-called Mitchell and Long Adits. Little gold was produced and mining ceased by 1893. In 1905, James Mitchell began work at the site, founding the Ogofau Proprietary Gold Mining Company Ltd in 1906. After limited successes, the lease was transferred in 1909 to Cothy Mines Ltd which continued to employ Mitchell as the mine manager. Work now concentrated on sinking a 30m (100ft) deep shaft in the Ogofau Pit, from which exploration of the mineralization was possible. Extensive problems with flooding halted operations in 1910, and the company folded in 1912.

(3) Mining in the 1930s. Britain's abandonment of the gold standard in 1931 led to renewed interest, culminating in the formation of Roman Deep Ltd in 1933. Initial exploration results encouraged the creation of a new company, Roman Deep Holdings Ltd, in 1935, with the necessary capital both to extend the 1909 shaft to 152m (500ft) and to drive the necessary exploratory cross-cuts at various levels. By mid 1936, satisfactory yields were being reported from the development ore shipped to Hamburg. Thus, in 1937, the parent company formed a subsidiary operating company, British Goldfields (No.1) Ltd, and sufficient capital was raised to purchase the relevant surface plant and mining equipment. Construction work was also begun on a new processing mill, which became operational in early 1938, and on the tramway connecting it to the shaft. Shortages of capital, however, coupled with various other problems, brought milling to an end in late 1938, while pumping ceased in 1939.

11 References:

Annels, A.E. (1984) 'Exploration for Gold in Wales', Chartered Land Surveyor/Chartered Mineral Surveyor RICS, 2, No. 12, 645-655.
Annels, A.E. and Burnham, B.C. (1986) The Dolaucothi Gold Mines, University College Cardiff, University of Wales, 2nd Edition.
Austin, D. and Burnham, B.C. (1984) 'A new milling and processing complex at Dolaucothi: some recent fieldwork results', Bulletin of Board of Celtic Studies, 31, 304-313.
Bick, D. E. (1988) 'An ancient leat near Dolaucothi Goldmine', Archaeology in Wales, 28, 20-21.
Boon, G.C. and Williams, C. (1966) 'The Dolaucothi drainage wheel', Journal of Roman Studies, 56, 122-127.
Davies, O. (1936) 'Finds at Dolaucothy', Archaeologia Cambrensis, 19, 51-57.
Hall, G.W. (1971) Metal Mines of Southern Wales, Jennings, Gloucester, 39-48.

Healy, J.F. (1978) Mining and Metallurgy in the Greek and Roman World, Thames and Hudson, London.

Holman, B.W. (1911) 'Gold Deposits of Cothy, South Wales', Mining Magazine, May, 374-378.

Jones, G.D.B, Blakey, I.J. and Macpherson, E.C.F. (1960) 'Dolaucothi: the Roman Aqueduct', Bulletin of Board of Celtic Studies, 19, 71-84.

Lewis, P.R. (1977) The Ogofau Roman Gold Mines at Dolaucothi, The National Trust Yearbook 1976-77.

Lewis, P.R. and Jones, G.D.B. (1969) 'The Dolaucothi Gold Mines - I: the surface evidence', Antiquaries Journal, 49, 244-272.

Murchison, R. (1839) The Silurian System, London 367-369.

National Trust (1986) Dolaucothi Gold Mines. Souvenir Booklet, Llandeilo.

Nelson, T.R.H (1944) 'Gold Mining in South Wales', Mine and Quarry Engineering, Jan. pp.3-10, Feb. pp. 33-38, Mar. pp. 55-61.

Smyth, W.W. (1846) 'Note on the Gogofau or Ogofau Mine near Pumpsaint, Carmarthenshire', Memoirs of the Geological Survey, 480-484.

12 Adjacent sites of interest:

NANTYMWYN LEADMINE, in Nant-y-Bai and around Rhandirmwyn village. NGRs SN 787446, SN 784446, SN 782437.

Text provided by Dr Barry C. Burnham, Archaeology Unit, St David's University College, Lampeter, Dyfed SA48 7ED.

SITE No. 5
Bordesley Abbey

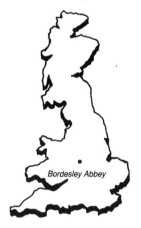

Bordesley Abbey

1 Name of site:

Bordesley Abbey, Redditch, Worcestershire.   Scheduled Ancient
Monument.

2 Metal:

Iron and possibly lead and copper.

3 General history of area:

The Cistercian Abbey of St Mary at Bordesley enjoyed 400 years of active
life from 1138 to 1538; aspects of its buildings, both religious and secular,
and technology have been explored in recent excavations, as have the
prehistoric, Roman and other pre-monastic exploitation of the Arrow
valley. Three phases of landscape may be discerned: pre-monastic, monastic
and post-Dissolution.

The end of the abbey as an active institution led to extensive destruction
for monetary gain, including quarrying of the ruins. However, the chapel
of St Stephen, the gateway chapel, continued in use until 1805 when its
function was transferred to a new church at the centre of Redditch town.
The town's principal recent industry, that of needle making, is represented
by an eighteenth century needle mill close by, which was restored and
opened as a museum of needle making, with a working waterwheel.

Bordesley Abbey. The water channel of the first mill. Head race at the top, wheel pit in the middle and tail race at the bottom of the photograph.
Copyright © Dr G. Astill.

In the 1960s the quiet rural area exploded into activity with the inception of the new town of Redditch. It was proposed to level the Abbey site and create playing fields, etc. However the entire precinct has now been preserved.

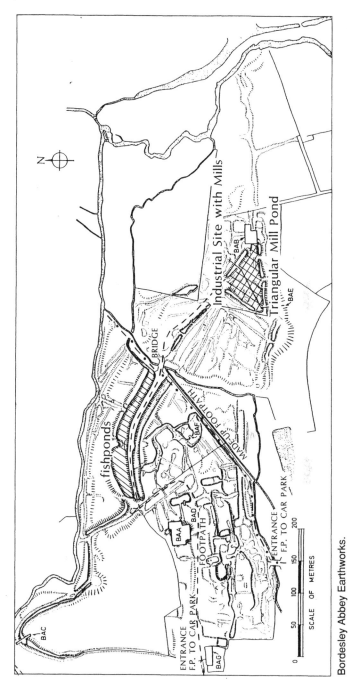

Bordesley Abbey Earthworks.
Copyright © Dr G. Astill & University of Reading, Department of Archaeology

4 Remains and dating:

The industrial site includes a mill, headrace, wheelpit and tailrace, dated to the late 12th century. Waterlogging ensured very good preservation of wood, leather and metal. Remains of four superimposed mills from late 12th century to late 14th century were identified connected with metalworking. The mill timbers have been removed for freeze drying and should be available for inspection. There are impressive earthworks, including remains of the triangular mill pond.

5 Location:

NGR SP 045857 2km from Redditch town centre.

6 Accessibility:

Open to the public.

7 Ownership:

Redditch Borough Council-Amenity Park.

8 Permission required to visit:

None required.

9 Sketch plan of site:

See page 47.

10 History of working at the site:

The first mill was constructed in the late 12th century and in the area of the leat evidence was found for the clearance of the ground prior to building operations. Remains of the head and tailrace and mill building were found.

The timbers forming the sluice gates and bank revetting of the headrace appear to have been part of the original mill and remained unchanged throughout the time the site was occupied. No timbers survived from the original wheel frame. The massive timber tailrace was excavated and lifted. One of the major timber seatings of the bottom of the tailrace showed clear signs of re-use. There is thus a possibility of an even earlier tailrace which was dismantled and re-used. The first mill building was earthfast, and the most substantial part of the building (the southern half) had uprights which were founded in post pits 0.4m square and faced the

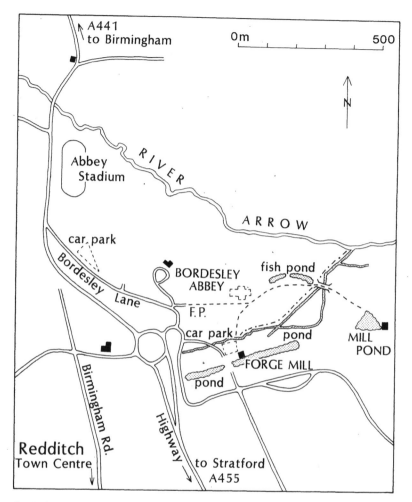

Bordesley Abbey. General plan of area.
Courtesy of Dr G. Astill

wheelpit. The northern part of the building was less sturdy and appears to have been more like a lean-to. Charcoal lined pits, interpreted as hearths, have been located in that part closest to the wheelpit. It seems that the earliest mill was used to provide power for metalworking.

At some time in the 13th century the post building was replaced by a padstone structure. It was of a similar character to that already excavated. The major uprights of the early mill were removed and the pits filled with clay, iron slag, charcoal and daub. Three other mills, superimposed on the twelfth-century mill, were also found. Well preserved evidence for head-race, wheel pit and tail race was found for each phase as well as padstone buildings. The mills contained a series of pitched tile hearths and are therefore interpreted as metal-working mills. The site was abandoned in the late 14th century.

11 References:

Astill, G. (1989) 'Monastic research designs: Bordesley Abbey and the Arrow Valley' in Gilchrist, R. and Mytum, H.(eds.) The Archaeology of Rural Monasteries, BAR British Series 203, BAR, Oxford.

12 Adjacent sites of interest:

The restored NEEDLE MILL and MUSEUM are part of the Redditch Borough Council Amenity Park.

Information from Prof. Philip Rahtz and Dr Grenville Astill, Department of Archaeology, University of Reading, Whiteknights, PO Box 218, Reading RG6 2AA.

SITE No. 6
Cwmystwyth

Cwmystwyth

1 Name of site:

Cwmystwyth, Copa Hill, Comet Lode Opencast, Dyfed. Scheduled Ancient Monument, under the protection of Cadw Welsh Historic Monuments. N.B. 'COPA' means Upper in Welsh. So Copa Hill is Upper Hill not Copper Hill. There is no association with copper although the site is now referred to in literature as 'Copper Hill'.

2 Metal:

Copper in antiquity: lead/zinc from medieval times.

3 General history of area:

The early history of Cwmystwyth Mine is a period of great speculation on very little information. The first traces of man in Ceredigion, formerly Cardiganshire, appear during the Neolithic or New Stone Age and there has been a total absence of the remains of this era in the vicinity of the Upper Ystwyth valley. Until recently this has largely been true of the later periods of prehistory - and there is an apparent deficiency of archaeological sites in the immediate area. However, the lack of visible and identifiable remains is perhaps not surprising in view of the thick blanket cover of peat present on the hills and the lack of any systematic field walking coverage in the region. Of the 200 mines of any consequence in mid Wales, the majority are pre 1800 in origin though later activity has destroyed much of the evidence.

It is also worth noting that many of the more ancient mines have an above average gold content in the waste rock, it would not be presumptious therefore to assume that the grades which were mined yielded a small quantity of that metal. Gold and Electrum exist in tiny grains (50 microns or less) in the sulphide ores but the yield is generally in the region of half a gram per tonne. In cupiferous gossan the yield could be increased by a factor of 20 or more and form a small workable deposit; secondary copper minerals in the gossan are also more easily smelted than the unoxidized chalcopyrite. It is, therefore, this type of deposit which would have attracted early exploitation and may possibly have been all worked out by the time the Romans invaded.

Traces of the Roman occupation within the general area are more common. Four miles to the NNE lies the fortlet of Cae Gear, whilst eight miles downstream, alongside the River Ystwyth, lies the fortlet of Trawscoed and the main north-south Roman Road of Wales known as Sarn Helen. To the west of Cwmystwyth, seven miles distant lies the fortlet of Esgairperfedd.

Thirty miles to the south lies Dolaucothi Mine which has been established as having been worked by the Romans, but this only proves that the technology of organized mining and complex ore dressing existed in the area before AD 100.

A large area of old dumps on the west face of Copa Hill associated with quite rudimentary workings for lead on the Kingside Lode have been referred to as the 'Roman Dumps' although there is no evidence at all of surface remains of such antiquity. Many hand-dressing mortars, or 'bucking' stones, are associated with collapsed drystone wall shelters on these tips but crushing was more likely effected with iron hammers. Indeed many of the structures are built from mine waste that exhibit the occasional gunpowder shot holes and, therefore, an eighteenth-century or even nineteenth-century date is probable for some of the workings, although there may well be evidence here for mining earlier than this.

On the southwest slope of Copa Hill over 60 acres of topsoil have been removed by hushing, a practice used greatly by the Romans at their Spanish gold mines and also on a smaller scale at Dolaucothi. But this provides no terminus ante quem and written records show this type of hydraulic mining was being used on the site as late as 1785 (S.J. Hughes). The majority of hushing channels on this part of the hill appear to radiate out from a dam of probable eighteenth-century date and, more than likely, had a prospecting function.

In conclusion, the possibility remains of Roman workings for lead at Cwmystwyth, and on Copa Hill, but it must be borne in mind that the lead here was notably silver-poor in comparison with much of the mid Wales area and one might well consider a rather more favourable choice elsewhere, given the extent of their occupation and interests in metals. However, it is the opinion of Hughes (1981) that of the estimated 250000

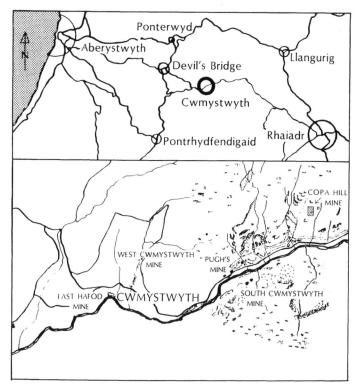

Cwmystwyth. General plan of area.
Copyright © S.J.S. Hughes

tons of development rock and gangue on Copa Hill (presumably the Kingside Dumps) at least 90000 tons may have been produced during the Roman period.

It would seem unlikely that any exploitation of the mines took place immediately after the withdrawal of the Romans from Wales and none of the contemporary writings before the 11th century suggest otherwise.

It would seem likely that the next lead miners, or probably the first, were the Cistercian monks from the Abbey of Strata Florida, or else persons appointed by them as miners.

The first monks to arrive in the area settled about 8 miles to the south-west of the mine at an Abbey known as Hen Fynachlog. The generally accepted date for the completion of this first Abbey is 1164 with a monk named David being appointed as the first Abbot. Due to philanthropy and, possibly, a natural disaster, a second Abbey was built and completed in 1202. This Abbey was one of, if not the grandest in Wales. It retained the

53

Looking north towards ancient opencast at the top of the Coment Lode.
The lode here cuts the brow of the hill and is visible on the skyline.
Copyright © S. Timberlake

name of Strata Florida and had as its first abbot Sissilus, who was the abbot
of Hen Fynachlog when it was visited by Giraldus Cambrensia and
Archbishop Baldwin in the year 1188.

It is in the remains of the new Abbey that the links with mining become
evident. The monks were obviously capable engineers in that they built a
leat system for their corn mill and used part of the tailrace to flush their
underground sewers. Their fresh water was brought to the Abbey by
means of a 4in diameter lead pipe. Stephen Williams, in his account of the
Abbey, relates how part of a smelting furnace was found close by, and
since smelting slag was strewn about the grounds, and on the strength of
an assay done on one of the finds from the 1887 excavation, concluded that
the monks practised cupellation due to the low silver content of the
metallic lead.

During the 13th and 14th centuries the boundaries of the mine fell
within three divisions: (i) the Parish or Lordship of Llanbadarn Fawr; (ii)
the Grange of Cwmystwyth; (iii) the Tenement of Briwnant. Categories (ii)
and (iii) at this date can make reference to no site other than Cwmystwyth;
the Lordship of Llanbadarn Fawr however covers the northern half of

Ceredigion and the references to mines at Llanbadarn could refer to any of a dozen or more sites.

There is one reference to a single, but important lead mine in Llanbadarn in 1282, and to a mine in West Wales, 1301, with 'plenty of ore'. There were two Inspectors of Silver at Cardigan in 1340. The 14th century is poorly represented and only a single reference in 1485 can be found that would encompass the Cwmystwyth Mines as well as the other north Ceredigion mines.

Summing up the evidence for mining before 1500, it is seen that there are no contemporary reports or accounts of mining at Cwmystwyth alone.

Whatever work - if any - was done between 1300 and 1500 amounted to very little. Probably less than 200 tons of ore as a maximum were raised between these dates. It would certainly have been advantageous for the first Cistercians to work the mines if only to supply some of the lead used in roofing the abbey and the fabrication of pipes and the latticework around the windows. As the dimensions of the abbey are known, it can be reasoned that the weight of lead needed to roof the structure with $^3/_{16}$ths of an inch sheet gives a figure of 70 tons.

As regards specific documentary evidence for mining at Cwmystwyth there is nothing until the beginning of the 16th century at which point a lease was issued for Cwmystwyth Mine to Rhys and David ab Ievan ab Hywel from Abbot Richard Talley of Strata Florida. There is some uncertainty about the exact date but it would appear to be around 1535. Of interest also is the account of the visit of Leland, King Henry VIII's antiquary, about this period and his journey down the valley as he approached Copa Hill on the old monastic road from Rhayader:

> 'About the middle of this Wstwith valley that I ride in being as I guess three miles in length I saw on the right hand of the hillside Cloth Moyne [Anglice: Cloth or Clodd = Mine : Moyne or Mwyn = lead ore] where hath been a great digging for Leade, the smelting whereof hath destroid the woodes that sometimes grew plentifully thereabout.
>
> I heard a marvellous tale of a crow fed by a digger there that took away his feeder's purse and while the digger followed the crow for his purse the residue of his fellows were oppressed in the pit with a ruin.'

These comments of Leland are the best circumstantial evidence for lead mining having been established here prior to 1500.

Of probable greatest antiquity on the site (Copa Hill) is a primitive opencast working situated at the top of the Comet Lode where it outcrops on the brow of the hill overlooking the valley. The lode here carries (relatively) appreciable amounts of chalcopyrite ($CuFeS_2$) ore, it being the only obvious occurrence of copper at surface in an otherwise lead-rich area. For up to 100m downslope of the infilled opencast are several eroded and since overgrown tips of shattered rock containing numerous bruised and split pebble hammers.

The occurrence of stone tools here was first noted as long ago as 1848 and in the mid-1930s the site was examined by Oliver Davies as part of the investigations of the British Association Committee set up to examine the evidence for early mining in Wales. Three of the tips were partially sectioned by Davies, who concluded:

'We cannot do more at present than guess at the date of the Comet Lode opencast. I am inclined to think that the hammer querns are old Celtic, approximately contemporary with the Roman period, though surviving after it'.

The opencast area and tips have since been the scene of recent survey work and archaeological excavations carried out by Simon Timberlake and the Early Mines Research Group in 1986 and 1989.

4 Remains and dating:

(1)*Opencast and tips with stone tools.* A section through one of these tips cut in 1986 provided burnt wood from fire setting activities from which three carbon-14 dates have since been obtained. These calibrated suggest a date of c.1500 BC indicating mining for copper here in the Early Middle Bronze Age. A channel of likely artificial origin buried beneath one of the tips (possibly with an ore washing or hushing function) is probably of the same period.

Deep excavations within the opencast area during the summer of 1989 revealed a great overburden of later deposits sealing the early tip and sides of the quarry, including natural silt and peat accumulation, remains of a water hushing dam and reservoir and spoil from an eighteenth century shaft. Against the north wall of the opencast at depth was discovered a fire-set gallery with stone tool marks in the roof, of likely prehistoric age. Above and leading into the opencast is the outline of an early leat of uncertain age and function but possibly connected with a later hushing use of the quarry. The outlines of rectangular structures downslope along the 400m contour appear to be much later features unconnected with any mining activity.

(2)*Tools.* The numerous pebble hammers are not grooved although some are slightly notched - they were probably hafted. (N.B. The cobble hammers are similar in appearance to mining tools found associated with Early Bronze Age copper mines in Ireland and many EBA and Chalcolithic sites on the Continent. Similarly there are close parallels with New World sites such as early Amerindian mining for native copper on the Keewanaw Peninsular in the Lake Superior Region, USA and pre-Columbian mining at Chuquicamata in Chile.)

A large ore dressing stone ('saddle quern' type) was found on site. Its use may have been contemporary with that of mining hammers. Remains of red deer antlers were also found within the tips (?possible tools). Examples of stone hammers may be seen on site. Other finds are within the National Museum of Wales.

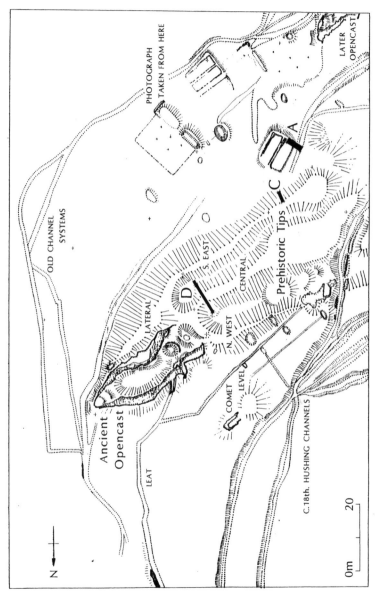

Copa Hill. Comet Lode Opencast.
Copyright © S. Timberlake

5 Location:

NGR SN 816 756 on western slopes of Copa Hill at an altitude of 400m OD, overlooking Cwmystwyth valley at a distance of 2km from Cwmystwyth village.

6 Accessibility:

Openland with tough climb to top of hill.

7 Ownership:

Crown Estate Commissioners Mr J. Raw, Ty Llwyd Farm.

8 Permission required to visit:

None required, but a courtesy call at Ty Llwyd Farm approximately 1km east along valley from site, if a large party is involved.

9 Sketch plan of site:

Shown on p. 57.

10 History of working at the site:

See section 3 and S. Timberlake (1987).

11 References:

Timberlake, S. (1987) 'An archaeological investigation of early mineworkings on Copa Hill, Cwmystwyth', Archaeology in Wales, 27, 18-20.
Timberlake, S. (1988) 'An archaeological investigation of early mineworkings on Copa Hill, Cwmystwyth,: new evidence of prehistoric mining', Proceedings of the Prehistoric Society, 54, 329-333.
Timberlake, S. (1988) 'Excavations at Parys Mountain and Nantyreira Mines', Archaeology in Wales, 28, 11-17.
Pickin, J. and Timberlake, S. (1988) 'Stone Hammers and firesetting: A preliminary experiment at Cwmystwyth Mine', Bulletin of Peak District Mines Hist. Society, 10(3).
Davies, O. (1935) Roman Mines in Europe, Clarendon Press, Oxford, pp.154-166.
Davies, O. (1947) 'Cwmystwyth Mines', Archaeologia Cambrenisis, 99, 57-63.
Hughes, S.J.H. (1981) 'The Cwmystwyth Mines' British Mining No. 17. A monograph of N.M.R.S.

Morris, Lewis (1744) 'An Account of Lead and Silver Mines in Cwmmwd y Perveth', p.46. National Library of Wales, Mss.

12 Adjacent sites of interest:

COPA HILL, CWMYSTWYTH. West slopes of hill north (2-300m) of the Comet Lode working: Kingside or 'Roman' Dumps. Seventeenth to early nineteenth-century workings on a primitive scale for lead: large areas of tips with drystone wall foundations of huts for hand ore crushing with much evidence of 'bucking stones' for ore grinding. Also nearby eighteenth-century hushing dam.

CWMYSTWYTH KINGSIDE MINE. Nineteenth to early twentieth-century mining for lead and zinc. Level Fawr entrance behind remains of corrugated sheet covered dressing mill (demolished August 1989). Mechanized mining operation.

GRAIG FAWR, CWYMYSTWYTH. Large opencast on hilltop above Level Fawr workings. Visible from road as large crag: c. eighteenth-century gunpowder working. Adits lead into it from top of Nant Watcyn stream. N.B. Henry's Roman adit - hand picked 'coffin level' is probably of eighteenth century date.

NANTYREIRA MINE. On the eastern slopes of Plymlimon, west of Llanidloes, approximately 32km from Cwmystwyth by road at NGR SN 826874, 500m OD - a primitive mine of early Bronze Age date (two radiocarbon dates), for copper, possibly re-worked in 19th century.

DAREN MINE AND HILLFORT. NGR SN 678831 near Penrhyncoch, Aberystwyth. Very old open cut glancing Iron Age hillfort over top of hill: lead/copper vein of unknown age. Twll-y-mwym at one end of vein at NGR SN 682834 locally rich in copper possibly a Bronze Age working. Stone hammers with much charcoal found inside old workings broken into in 1740s (ref. Lewis Morris). A replica of a primitive spade found at Daren, in old workings in the 19th century, is in the Ceredigion Museum in Aberysthwyth. An ancient smelting mill for Daren lies some two miles to the west at Felin Hen, i.e. the old mill. It straddles the Roman road known as Sarn Helen about a mile north of Capel Bangor fort.

ERGLODD MINE AT TALYBONT. NGR SN 659904 Nineteenth-century lead mine with much earlier opencast possibly for copper: stone hammers on surface, possibly Bronze Age.

Information from Simon Timberlake, 12 York Street, Cambridge, CB1 2PY and Simon Hughes, Leri Mills, Talybont, nr Aberystwyth ,Dyfed, SY24 5ED.

SITE No. 7
Bryn y Castell

1 Name of site:

Bryn y Castell Hillfort, Ffestiniog, Gwynedd. Scheduled Ancient Monument.

2 Metal:

Iron.

3 General history of area:

Snowdonia is an area rich in prehistoric remains, which include some 90 hillforts and a very large number of prehistoric and Romano-British settlements. Few of the hillforts have been excavated and have rarely produced significant finds. It is assumed that most ceased to be used after the Roman conquest of the area c.80 AD, but a small number of forts have evidence of occupation during the Roman period. There are several important finds of late-prehistoric bronze and iron objects from the area, including the Capel Garmon fire-dog and the Trawsfynydd tankard.

4 Remains and dating:

Bryn y Castell has been completely excavated, and restored, by the National Park. The hillfort had low drystone ramparts enclosing a pear-

Bryn y Castell.
Copyright © Peter Crew

shaped area some 40m by 25m with two entrances. Smelting furnaces were found outside to the north and inside at the southern end. Internal buildings included two stake-wall round houses and a large stone founded round house, with four internal roof support posts. This latter structure was re-built in the late prehistoric period in a 'snail' shape and was shown to have been used as a smithy. A rectangular platform at the southern end of the hillfort may also have been a smithy. Radio-carbon and archaeomagnetic dates show that the hillfort was in use during the last century BC and evidence indicates ironworking throughout this period, intensifying about the time of Agricola's AD 80 campaign.

A small external structure to the north was used exclusively for iron-working, containing some 30 furnaces/hearths. A large 650kg external dump of slag included a stone anvil and hammerstones.

In total 1200kg of iron-working residues were found. Subsequent experimental work has shown that this is probably the product of smelt-ing, bloom smithing and smithing and it is possible that iron products were traded. Other finds include decorated glass armlets, game boards, gaming pieces and a wide range of stone artefacts. These can be inspected at Plas Tan y Bwlch. The stone anvil remains on site.

5 Location:

NGR SH 728429 on the summit of small steep-sided rocky knoll 380m above sea level, Ffestiniog, Gwynedd. Track from B4391 Bala-Ffestiniog road.

6 Accessibility:

Public footpath (partly following Roman Road) passes near site.

7 Ownership:

Crownland Tenant: H. Hughes, Teilau Bach, Ffestiniog.

8 Permission required to visit:

None required, though courtsey call to Mr Hughes desirable for large parties. This can be arranged through Snowdonia National Park Centre at Plas Tan y Bwlch, Ffestiniog.

9 Sketch plan of site:

Shown on page 64.

10 History of working at the site:

See section 4 on page 61.

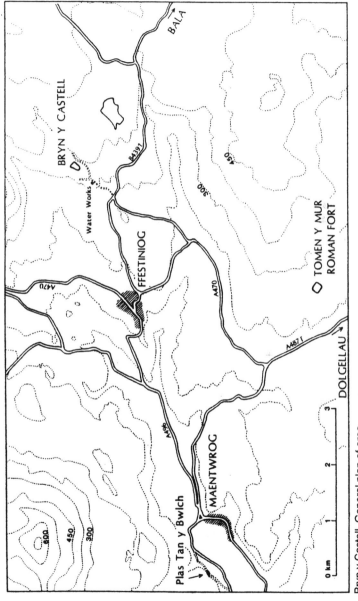

Bryn y Castell. General plan of area.
Copyright © P. Crew

63

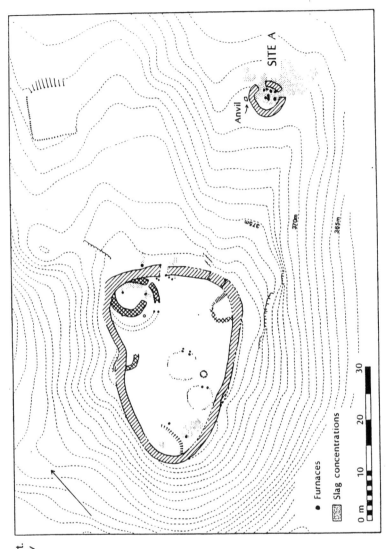

SITE A

Anvil

375m
370m
365m

• Furnaces
▦ Slag concentrations

0 m   10   20   30

Bryn y Castell Hillfort.
Copyright © P. Crew

64

11 References:

Crew, P. (1984) 'Bryn y Castell Hillfort - a Late Prehistoric Iron Working Settlement in North West Wales' Proceedings of the Symposium of UISPP Comite pour la Siderurgie Ancienne, The Crafts of the Blacksmith, 91-100, Ulster Museum, Belfast.

12 Adjacent sites of interest:

CRAWCWELLT WEST. Prehistoric iron-working site, partly excavated and consolidated (NGR SH 686308).

COED Y BRENIN MEDIEVAL BLOOMERIES (NGR SH 72 27).

DOL Y CLOCHYDD. Sixteenth-century blast furnace, (excavated and consolidated) NGR SH 734219.

DOLGUN. Eighteenth-century blast furnace, (excavated and consolidated) NGR SH 752187.

Information from Peter Crew, Snowdonia National Park Study Centre, Maentwrog, Blaenau Ffestiniog, Gwynedd LL41 3YU (Tel: 0766 85 324).

SITE No. 8
Alderley Edge

Alderley Edge

1 Name of site:

Alderley Edge, Cheshire.

2 Metal:

Copper.

3 General history of area:

Alderley Edge is a prominent escarpment of Triassic sandstone containing locally intensive deposits of copper disseminated in the host rock. The area has been worked since the late 17th century for copper and, to a lesser extent, lead and cobalt but the occurrence of stone hammers at some of the mines suggests the possibility of earlier, perhaps prehistoric, exploitation.

The earliest documented mining phase dates from 1693 when workings were centred on the surface exposures of carbonate ores in the north-eastern part of the Edge and at the Engine Vein Mine. Later workings were concentrated further to the west where the extensive West and Wood Mines were opened in the 1850s. Large-scale mining had ceased by the early 1880s but sporadic working continued until 1919.

The Engine Vein, Alderley Edge Mines. This photograph shows the pecked surfaces as well as the depth of the pits.
Copyright © D. Gale.

There is no documentary or archaeological evidence for medieval or Roman mining. Alderley Edge earned its reputation as a 'prehistoric' mining site from a spate of antiquarian activity in the late 19th and early 20th centuries. The Edge was conceived as a site of Bronze Age copper mining by William Boyd-Dawkins after his visit to the Brindlow Levels in 1874 being worked by the Alderley Edge Mining Company. He noticed grooved stone hammers in recently exposed hollows. Over 100 of these from the Engine Vein, Stormy Point and Brindlow sites were divided into three types:

(1) Hammers with transverse grooves
(2) as type (1) with a lateral groove
(3) Wedges with a flat head and groove.

An oak shovel which had been very roughly used, was found at Brindlow in 1878.

Roeder carried out a more systematic search over 1901-1905 and found four finds spots at Windmill Wood, Dickens Wood (Stormy Point), Mottram St Andrew and Engine Vein. These tools, examples of which can be seen in the Manchester Musuem and Oxford's Ashmolean Museum, are massive stone cobbles which are assumed to have been used for hard rock mining on a fire-set face and in ore dressing. Many have a central groove pecked around the circumference to hold a flexible handle or withy. None has been found in a secure archaeological context and so their date is problematic but they are very similar to grooved hammers found in known Bronze Age mines in Spain and Yugoslavia. They also have parallels with the grooved hammers found in pre-Columbian copper mines in Chile and Lake Superior where such tools are an indicator of very early and technologically uncomplex mining. Bronze Age mining in the British Isles is now a well attested fact with three mines in Wales and Ireland having produced a series of radiocarbon dates ranging from 1600 cal BC to 1000 cal BC. Stone hammers occur at all these sites but none, however, are really comparable with those from Alderley Edge. Most are simple, unmodified cobbles, and a few have rough notches, or slight grooves while the Alderley examples are distinguished by their deep and often multiple grooves and the extreme range in shape and size. Do these differences reflect nothing more than a local peculiarity within the general stone-using mining technology or do they indicate a quite different date from those at the Welsh and Irish mines? There is an obvious need for further research on Alderley's apparently unique mining tools and the answer must lie in future excavation aimed at obtaining secure radiocarbon dates.

4 Remains and dating:

There are remains of post-medieval mining in many parts of Alderley Edge and a number of the mines have been opened up and made accessible underground through the dedicated work of the Derbyshire Caving Club. At two mines, however, there are isolated traces of earlier mining which relates to working with stone tools:

(1) Engine Vein. This is an impressive opencast working where copper deposits have been mined from both sides of a major fault. The surface remains are predominantly 18th and 19th century but these operations have exposed and truncated a series of five earlier circular pits. They are most obvious on the opencast northern face and clearly show the pock-marked traces of hammering made with stone tools. These are quite distinct from the oblique and pointed marks left by metal tools and can be closely examined in the semi-circular pit exposed at the western end of the mine.

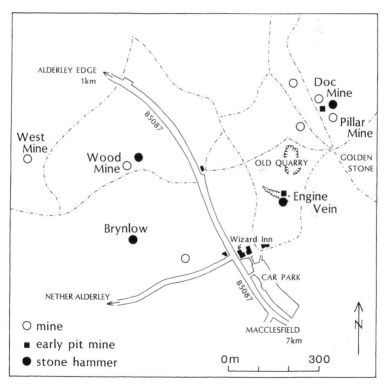

Alderley Edge Mines.
Courtesy of The National Trust

69

(2) Stormy Point. A series of conjoined and truncated pits are exposed in the face between the western side of Pillar Mine and the nineteenth century incline adit to Doc Mine. They exhibit the same stone tool batter marks as those in Engine Vein and may have once followed the bedded copper deposit running over the top of Pillar Mine. Stone hammers have occasionally been found in the spoil heaps running downslope. Pillar Mine itself is undated but its massive cave-like mouth is unique on the Edge and the curving profile of the entrance section is suggestive of fire-setting.

5 Location:

NGR SJ 858774 (centred on Engine Vein). 4.5 miles NW of Macclesfield and 1.25 miles SE of village of Alderley Edge on both sides of the B5087.

6 Accessibility:

Open access at all reasonable times. There is a large National Trust car park at the side of the Wizard Hotel.

7 Ownership:

National Trust.

8 Permission to visit:

Permission to visit is not required but underground access is controlled by the National Trust through the Derbyshire Caving Club with whom prior arrangments must be made, through:

Nigel Dibben, Chairman, 62 Middlewich Road, Holmes Chapel, Cheshire, CW4 7EB. Tel: Day 061-775-2644; evening 0477 34778.

9 Sketch plan of the site:

Shown on page 69.

10 History of working at site:

See section 3.

11 References:

Boyd-Dawkins, W. (1875) 'On the Stone Mining Tools from Alderley Edge', Journal Anthropological Inst. GB and Ireland, 5, 2-5.

Roeder, C. (1901) 'Prehistoric and subsequent Mining at Alderley Edge', Trans. Lancs. and Cheshire Antiquarian Society, 19, 77-118.
Carlon, C.J. (1979) The Alderley Edge Mines, John Sherratt & Son Ltd, Altrincham.
Warrington, G. (1981) 'The copper Mines of Alderley Edge and Mottram St Andrew, Cheshire', Journal Chester Archaeology Society, 64, 47-73.
Gale, David (1986) 'Recording an Elevation of a Copper Mining Face at Engine Vein, Alderley Edge'. Unpublished Thesis, Bradford University.
Findlow, Alan James, (1988) 'Descriptive account of the area known as Alderley Edge', Albion Press, Macclesfield.

12 Adjacent sites of interest:

Copper has also been worked at MOTTRAM ST ANDREW, 1.5 miles northeast of the Alderley Mines. Stone mining hammers have been recorded from this site but there are now no surface remains.

At BICKERTON, 16 miles southeast of Chester, copper has been worked in deposits similar to those at Alderley. Mining again dates from the late 17th century but the only surface remains are the stack and foundations of a nineteenth-century engine house.

Recent excavations at BEESTON CASTLE, 2 miles north, have uncovered the remains of a substantial Bronze Age copper workshop overlain by an Iron Age hillfort and the medieval castle. There has been some speculation that this workshop could have been connected with the exploitation of the Bickerton ore deposits.

Information from John Pickin, The Bowes Museum, Barnard Castle, Co. Durham, DL12 8NP and David Gale, 78 Hustler Street, Bradford, BD3 0PS.

SITE No. 9
Brantwood

1 Name of site:

Beck Leven Bloomery at Brantwood, Coniston, Cumbria.

2 Metal:

Iron.

3 General history of area:

Woodland and forest covered the fells while beneath the surface of the fells were considerable deposits of haematite iron ore (a very rich iron ore). The lakes provided fresh water, clay for furnace hearths and a means of transport. Here were all the requirements for iron ore smelting, the location being dictated by logistical problems. It was preferable to take the iron ore to the charcoal supply because of the number of pack horse loads required: the ratio is about 1:30. Sites alongside the major lakes filled the water requirements.

At Brantwood - and in fact along the Coniston Water shoreline - bloomeries were set up just in-shore, for the sake of security and secrecy. Iron ore came from mines in Low Furness and the Central Fells at the head of Little Langdale. It was unloaded at a pier on the lake shore fairly close to but not actually at the smelting site. The monks of Furness Abbey recorded their activities alongside the eastern shores of Coniston Water (then Thorstein's Mere).

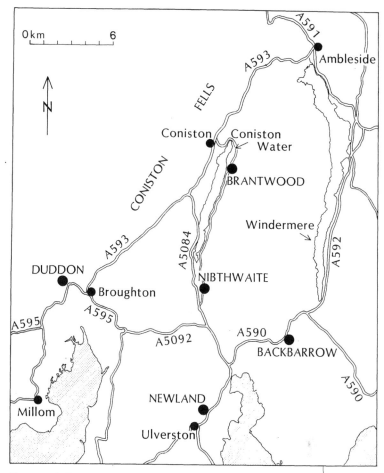

Brantwood. General plan of area.
Courtesy of M. Davies-Shiel

4 Remains and dating:

There is a large mound of bloomery slag visible from the roadside, but the hearth, or hearths, have not been identified or located. There is an interesting difference in characteristics of the slag remains. The site is first mentioned in the Furness Abbey records, the Furnace Coucher, in 1338 but it is likely that there had been smelting for some years previously.

5 Location:

NGR SD 310952. On the inland side of the lakeshore road running along the east shore of Coniston Water in the grounds of Brantwood House, - the home, between 1870 and 1901, of John Ruskin - $2\,^1/_2$ miles from Coniston.

6 Accessibility:

From the House (and car park) owned by the Brantwood Trust. Only open daily mid-March to mid-November then Wednesdays to Sundays (11.00-16.00) until mid-March.

7 Ownership:

Brantwood Trust.

8 Permission required to visit:

None required but access is via Brantwood House only.
    NO SLAG SAMPLES MAY BE REMOVED.

9 Sketch plan of site:

Shown on opposite page.

10 History of working at the site:

Recent survey work has proved that the grounds contain 75 charcoal pitsteads, four other bloomeries beside eight potash kilns and charcoal burner's huts. As mentioned above, some of the dumped slag pieces indicate previous molten state obtained from bloomsmithies, possibly by the use of water-powered bellows. It is likely that this development took place at the end of 15th and early 16th centuries. The Furnace Abbey records indicate the use of this process on the western shore of Windermere at Cunsey ('Les Smithies') around about 1500 to 1530. There is no proof that this process was in use at Beck Leven but the indications are that it probably was. A reconstruction of a bowl-type furnace is on the site.

11 References:

Furness Abbey Records and Coucher.

12 Adjacent sites of interest:

Eighteenth-century blast furnace remains at DUDDON (well restored) NGR SD 197883, BACKBARROW SD 355847, NEWLAND SD 299798 and NIBTHWAITE SD293898.

N.B. PERMISSION IS REQUIRED TO VISIT THESE SITES.

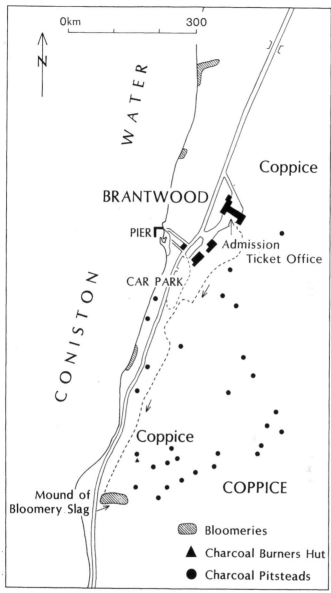

Beck Leven Bloomery at Brantwood.
Copyright © M. Davies-Shiel

N.B. PERMISSION IS REQUIRED TO VISIT THESE SITES.

Information from Michael Davies-Shiel, Micklethwaite, Annisgarth Park, Windermere, Cumbria, LA23 2HX.

SITE No. 10
Browngill

1 Name of site:

Browngill, Alston, Cumbria.

2 Metal:

Lead/silver.

3 General history of area:

The Alston Moor area (now in Cumbria but formerly largely in Cumberland) is criss-crossed with numerous veins of lead ore, some of which carry high values of silver. The origin of mining in the area is not known: the earliest documentary reference dates from the earliest surviving Pipe Roll (of 1130 AD) by which time the mines were known as 'the silver mines of Carlisle' and were worked largely for silver for the royal mints. It appears that later medieval mining was largely for lead, though the Crown retained control due to the silver output.

From the end of the 17th century (when the Crown monopoly of gold and silver mines was ended) the scale of mining on Alston Moor increased rapidly and for much of the 18th and 19th centuries the area was dominated by the London Lead Company, a Quaker-run company which was

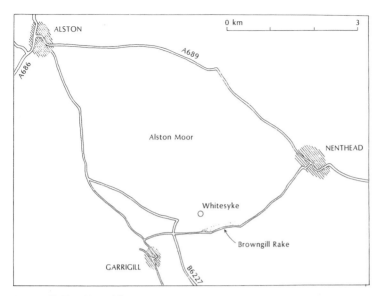

Browngill. The Alston Moor area.
Courtesy of The Ordnance Survey

one of the largest lead producers in Britain. The lead mining industry collapsed in the later 19th century, though zinc mining continued well into the 20th century.

Inevitably the lead veins were worked repeatedly over the centuries and Browngill was no exception. Finds of early coins (see below) suggest that it was one of the 'silver mines of Carlisle' but it is not possible to say which of the surviving features date from this period of exploitation.

4 Remains and dating:

At Browngill, and along the rake, is an impressive series of bell pits, rake workings, open-cut excavations and spoil heaps of hand washing debris. This was an argentiferous lead vein. Eleventh-century coins were found during nineteenth-century reworking along with heavily oxidized iron tools. The coins were pronounced to be those of William II (Rufus).

5 Location:

NGR NY 756421 to 765423 along half mile on north side of Nenthead-Garrigill road.

6 Accessibility:

On verges of public road. The rake extends on to private land, visible from the road but not closely accessible.

7 Ownership:

See section 6.

8 Permission required to visit:

None required, but private land not accessible.

9 Sketch plan of site:

None available but see map on page 77.

10 History of working at the site:

As indicated above, the whole area was intensively worked over nearly eight centuries.

11 References:

Moorhouse, F.J. (1942) 'Pre-Elizabethan mining law, with special reference to Alston Moor'. Transactions of the Cumberland and Westmorland Antiquarian and Archaeological Society, 45, 22-23.
Wallace, W. (1880) Alston Moor: its pastoral people: its mines and miners, Reprinted 1986, Davis Books, Newcastle-on-Tyne.
Walton, J. (1945) 'The medieval mines of Alston', Transactions of the Cumberland and Westmorland Antiquarian and Archaeological Society, 45, 22-33.

12 Adjacent sites of interest:

NENTHEAD MINES AND SMELTERS. Garrigill eighteenth-nineteenth century lead mining village: abundant post-medieval mining remains throughout the area.

Information from David Cranstone, 59A Dartmouth Avenue, Low Fell, Gateshead, Tyne and Wear, NE9 6XA.

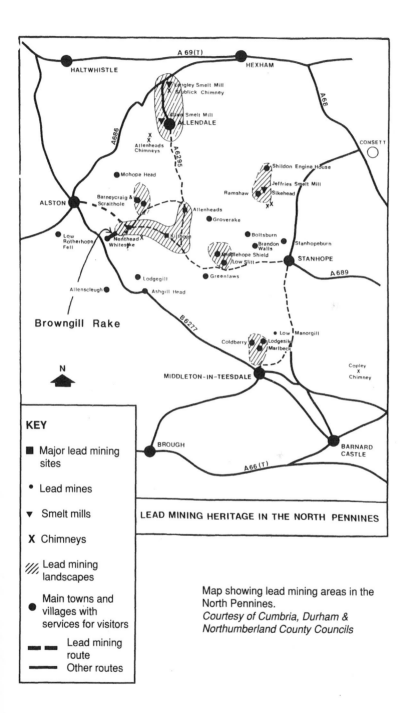

KEY

■ Major lead mining sites

• Lead mines

▼ Smelt mills

X Chimneys

/// Lead mining landscapes

● Main towns and villages with services for visitors

-- -- Lead mining route

—— Other routes

LEAD MINING HERITAGE IN THE NORTH PENNINES

Map showing lead mining areas in the North Pennines.
Courtesy of Cumbria, Durham & Northumberland County Councils

79

SITE No. 11
Peak District, Derbyshire

Peak District

1 Name of site:

Various sites around Matlock, Derbyshire.

2 Metal:

Lead.

3 General history of area:

The origin of lead mining in the Peak District cannot be accurately ascertained but the Romans definitely worked the ore, and 'pigs', or lumps, of metallic lead have been unearthed from time to time. The first was found in 1777 - this is positively dated to AD 117-138 - and now at least 29 are known. Most have inscriptions, in Latin, on one or more faces. The Derbyshire pigs are distinguished by the letters LVT or LVTVD, or in one case LVTVDARES, believed to refer to LUTUDARUM as the place of manufacture and this was supposed to have been near Wirksworth or Matlock but no positive evidence so far. In 1980, two sites were excavated about 3 miles west of Wirksworth, near Carsington (Current Archaeology 75, Vol VII, No. 4, 1981, 125-126) which revealed an extensive Romano-British settlement which is now presumed to be Lutudarum.

Ploughing, which revealed very extensive scattering of pottery and masonry material unfortunately virtually destroyed the sites. A sequence of occupation with timber structures and leadworking in the early to mid second century was identified followed by the erection of substantial stone buildings at end of the century. The site appears to have become agricultural by the 4th century. A further difficult problem is presented by the letters EX ARG - which might be inferred as relating to a source of silver - or from which the silver has been extracted. The Derbyshire ores were usually thought poor in silver.

The Romans probably extracted the lead ore in open workings along the outcrops of major veins, which could be between 40 and 60 ft wide at the surface and capable of opencast working to a fair depth: no actual workings have yet been identified.

After the Romans, the Saxons and the Danes continued mining. the Domesday Survey of 1086, lists seven lead works, which are thought to be smelters rather than mines. Until the end of the 13th century the Derbyshire miners' right to dig were based on customs and privileges. So in 1287 they petitioned the King, Edward I, to set down their customs and rights. An Inquisition was held at Ashbourne on the Saturday after Trinity Sunday 1288 to enquire into these claims. From this arose the QUO WARRANTO, now in the Public Record Office, which formed the nucleus of Derbyshire lead mining law for the next 600 years.

Part of the procedure was payment to the Barmaster (a Crown Official who dealt with all lead mining customs) of two 'freeing dishes' of ore, representing the initial payment due to the owners of the mineral duties. The Original dish presented to the miners by King Henry VIII in 1513 is still preserved in the Moot, at Wirksworth. Its volume is 14 Winchester pints.

From the early 17th century, the veins had been worked down to the water table and drainage channels, or soughs were driven through the ground at lower levels and drained into rivers two or three miles away.

Prior to 1600 the smelting of lead took place in 'boles' sited on west-facing hill tops, with wood as the fuel. The slag after the metal was cast was re-smelted with charcoal and more lead recovered. The prevailing wind was the draught. Around 1568-1580 a waterwheel operated bellows was applied to a furnace for greater and more efficient production.

The lead mining industry boomed from 1700 onwards for the next 150 years but from about 1850 the heyday was coming to an end and mines were starting to close, until mining ceased in the 1920s, with the exception of the Millclose Mine which worked until 1940. The last Peak District lead miner - C. H. Millington - died in 1968 aged 90.

This information is by the courtesy of the members of the Peak District Mines Historical Society: Trevor D. Ford and J. H. Rieuwerts (eds.) 1983 Lead Mining in the Peak District.

4 Remains and dating:

It is proposed that a visit to the Matlock area should commence at the Peak District Mining Museum (NGR SK295581): Matlock Bath, Derbyshire DE4 3PS Telephone 0629 583834 (open every day 11.00-16.00 except Christmas Day). Large numbers of mining and smelting artefacts and displays including a Romano-British smelting hearth, copies of the Bronze Dish and Roman lead ingot (other ingots are in the Industrial Museum, Derby and the Hull Museum).

5 and 6 Location and accessibility:

In Wirksworth are the following :-
(1) *Church* (NGR SK287539), in which is a stone carving of a medieval miner 'T'owd Man' carrying a corf and a pick. Access easy, prior notice helps.

(2) *Barmoot Hall* (NGR SK287541), in which is the 1512 Bronze dish for measuring lead ore (a contemporary copy is in the Science Museum and a modern copy is in the Peak District Museum). Access, via the Museum. There are many sites but there are difficulties in interpretation and sometimes access. Warm clothing and boots essential as is assistance by a member of the Peak District Mining Historical Society. The three chosen as most reasonable are:

(a) *Ramsley Moor* - Bolehill site on hilltop, but extant remains are nearby on NGR 279745 and downstream on the west bank. Public access is tolerated. Remains of water leats, and low mounds of lead slag. This part of site was most probably a washing place for slag and re-smelting in footblast 'ovens'. Late medieval. No remains of bole on nearby hill top.

(b) *Harland Moor* - Bolehill site at NGR SK30106815 (Trig. Point 367m). This site has burnt areas just below scarp top, on the scarp and about 100m north. Land is heavily poisoned. Reddened stones associated with burnt areas. Much lead slag scattered; 'store place' for partially burnt ore. Other bole sites within V-shaped perimeter of old pack-route. Twelfth-thirteenth-century dating is suggested (later boles were larger). Private land: access permission from Chatsworth Estate.

(c) *Beeley Moor* - Bole hills, along scarp edge near NGR SK282670, small every 4m. Slag and burnt stones mark sites. Smaller than Harland Edge (above)- less than 1 m diameter and therefore no later dating. Private land: rough going across near one mile - access permission through Chatsworth Estate.

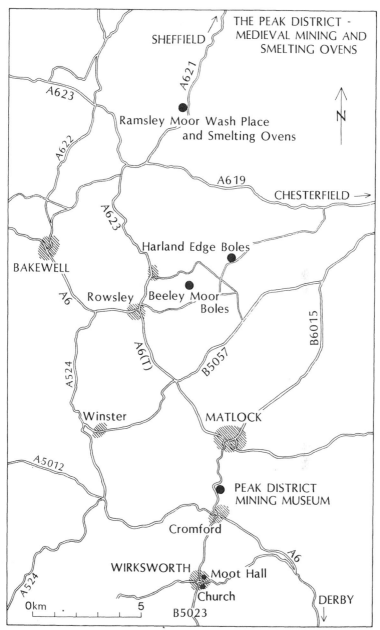

The Peak District. General plan of area showing mining and smelting ovens.
Courtesy of Dr L.M. Willies

7 Ownership:

Mainly Chatsworth Estates.

8 Permission required to visit:

From Chatsworth Estate via Peak District Mining Museum Office.

9 Sketch plan of site:

No site plan but sites are shown on page 83.

10 History of working at the site:

See section 4.

11 References:

Peak District Mines Historical Society Members 1983 (3rd Edition), Trevor D. Ford and J. H. Rieuwerts (eds.), Lead Mining in the Peak District, Peak District Planning Board, Bakewell.

Many others, too numerous to list: check with Peak District Mines Museum.

12 Adjacent sites of interest:

BONSALL LEYS (NGR SK268569). Romano-British settlement adjacent to leadworking - visible field system, settlement hollows, orthostat and other wall traces. Surface features of lead rake. Excellent seventeenth to eighteenth century small-scale workings visible at surface nearby. Access tolerated from public footpaths.

MIDDLETON MOOR (NRG SK268569). Similar to above.

ROYSTON GRANGE (NGR SK201568). Now an archaeological trail - call at Information Centre. Walls and field system of Roman mining visible. Also medieval (monastic) farming. Grange site - barn foundation.

STONE EDGE SMELT MILL. Near junction of B5057 and B6015 roads at approximately NGR SK342671. Tall square chimney at centre of site - remains of flues and furnace debris. First reference 1783; disused by 1852. Access can be arranged by the Peak District Mining Museum.

Information from Dr Lynn M. Willies. Peak District Mining Museum, Matlock Bath, Derbyshire DE4 3PS. Tel: 0629 583834.

SITE No. 12
Lostwithiel

Lostwithiel

1 Name of site:

Duchy Palace and Stannary, Lostwithiel, Cornwall.

2 Metal:

Tin.

3 General history of area:

The following information comes from Professor N.J.G Pounds's (1979) paper 'The Duchy Palace at Lostwithiel, Cornwall', Archaeol. J., 136, 203-217 and we are indebted to him for permission to quote therefrom.

About 1190 Robert de Cardinham granted a charter to his burgesses of Lostwithiel. In 1268-9 Isolda, his granddaughter, granted the borough and that part of the manor of Bodardle (which extended over a wide area of mid-Cornwall) 'to the east of the via regia which runs from Bodmin to Lostwithiel' together with Restormel Castle, to Richard, Earl of Cornwall and brother of King Henry III.

The territory then became the manor of Restormel. Shortly afterwards Earl Richard received 'the whole town of Lostwithiel and the Water of Fowey'. He died in 1272 and was succeeded in lands and title by his son, Edmund.

85

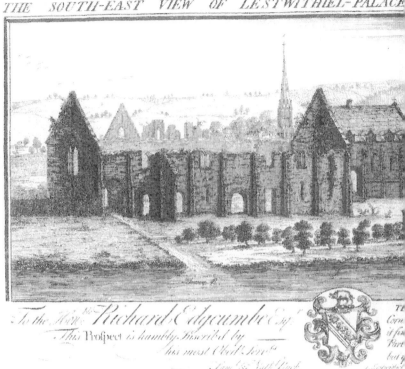

THE SOUTH-EAST VIEW OF LESTWITHIEL-PALACE

What the Earls Richard and Edmund had built up was a large estate in the West Country comprising the shrievalty of Cornwall, the castles of Launceston, Trematon, Tintagel and Restormel, the perquisites of the county and most of the hundred counts as well as more than 120 Knight's fees - and had acquired contiguous parcels of land around Lostwithiel.

It is probable that this area was attractive because it was central to the county, had access to shipping via the river Fowey and was close to the tin bearing lands of Bodmin Moor. They made it the centre of their administration and base of their authority within the county. The last Earl of Cornwall died without heir in 1336 and the rights of the earldom reverted to the Crown. In 1337, King Edward III conferred on his eldest son, Edward, known as the Black Prince, the title of Duke of Cornwall. The Duchy was endowed with the same lands and rights recently reverted to the Crown and they were made inalienable rights of the Monarch's first born son.

COUNTY OF CORNWALL.

Palace & Exchequer of the Earls and Dukes of
on Restormel Castle on ÿ adjacent Hill where
rkets Faires & examination of Burgesses for the
i nt the most Coynage of Tin only nth Sothers
ary & keeping of ÿ County Courts it self alone

The Buck engraving of the Duchy
Palace, 1734.
*Reproduced by courtesy of the
Cambridge University Library and
Prof N.J.G. Pounds.*

It appears that the construction of a cluster of buildings at Lostwithiel
- subsequently known as the Duchy Palace - commenced about 1289 under
Earl Edmund. An engraving of the Duchy Palace dated 1734, by Samuel
and Nathaniel Buck, exists which shows a great hall of eight bays with a
smaller hall to the north, known now as the Convocation Hall. Behind the
Great Hall is another building which is likely to have incorporated the
'blowing house' and the 'weighing house'. The functions of the Palace
were threefold. It was the administrative centre of the Duchy: it was the
venue of the County Court: and as Lostwithiel was a coinage town, the
earliest in Cornwall, serving the needs of the tinners roundabout, it was
the place where coinage took place, usually twice a year. The 'houses' in
which a corner (coin) of the tin ingot was melted and assayed lay behind
and to the west of the Great Hall. The Palace was never the residence of the
Earls or the Duke.

It appears that the upkeep of this fine range of buildings was onerous and there are frequent references to the poor conditions of the buildings. Nevertheless the Stannary Court, as well as the County Court, continued to meet there, though the business was declining in the reign of Queen Elizabeth I, and the gaol was in regular use. In 1644, the town was occupied by the Parliamentary forces who did such damage to the building that the Palace never recovered. It was ruinous in 1649-50, repaired in 1660-61, and again in 1681 but this did not prevent prisoners getting away. Early in the eighteenth century the significance of Lostwithiel was declining in favour of places in West Cornwall, the bulk of the tin being coined at Truro, Helston and Penzance. The Convocation of the Tinners met infrequently until 1752-3 after which it ceased to be summoned.

Sometime after 1734 a dwelling house was built in the remains of the northern end of the Great Hall, which served as the Cornish office of the Duchy until 1874. In 1878 the Convocation Hall was sold to the Freemasons who restored it, and still use it. The remainder of the Great Hall became a slaughter house, then a garage, and last an antique shop.

4 Remains and dating:

The surviving buildings comprise (1) the Convocation Hall; (2) the remains of the Great Hall; (3) an eighteenth-century building called the Stannary Gaol; and (4) ruinous buildings in the court to the west.

The Convocation Hall at the northern end of the Duchy Palace, Lostwithiel. *Copyright © John Blick.*

Medieval arch over the Cob brook (which now flows under the road) at the southern end of the Great hall of the Duchy Palace, Lostwithiel (extreme left arch in Buck engraving, shown on pages 86 and 87).
Copyright © John Blick

(1) *The Convocation Hall* is still in use. It is strongly buttressed and these buttresses are probably original thirteenth-century. In the late nineteenth-century dormer windows were removed, the northern gable replaced by a hipped roof and the upper floor taken out to form the present Freemason's Hall. It has a vaulted basement.

(2) *The Great Hall* consists of eight bays, again buttressed, with three entrances from the east. Much of the outer wall remains with a notable southern vaulted passage carrying a lane over the Cob Brook under the Hall.

(3) *The eighteenth-century buildings.* The Northern part of the Great Hall was taken down in the mid-eighteenth century and a dwelling house built. A shop front was added, but it was used for Stannary prisoners on the top floor.

(4) Stannary buildings. Those to west of the Great Hall contained the buildings used for assaying and weighing of tin but have been very much disturbed. An alley known as Smelting House Lane still exists. The area is difficult to interpret but remains of the slaughter house rather than the earlier primary use are indicated.

5 Location:

In the town of Lostwithiel parallel to the River Fowey. NGR SX 104596.

6 Accessibility:

Visible from exterior.

7 Ownership:

Various.

8 Permission required to visit:

No internal access.

9 Sketch plan of site:

No sketch plan but see opposite.

10 History of working at the site:

See sections 3 and 4.

11 References:

Pounds, Professor N.J.G. (1979) 'The Duchy Palace at Lostwithiel, Cornwall', Archaeological Journal, 136, 203-217.
Nicholls, D. de L. (1979) 'Lostwithiel'. Blackfords, Truro.
Sheppard, P. (1980) The Historical Towns of Cornwall- an Archaeological Survey. Published by Cornwall Committee for Rescue Archaeology, Truro.
Crown Copyright reserved.

12 Adjacent sites of interest:

THE MINIONS AREA - 4 miles north of Liskeard NGR SX 262712. Fine examples of nineteenth century Cornish Engine houses in a spectacular setting, e.g. Phoenix Mine.

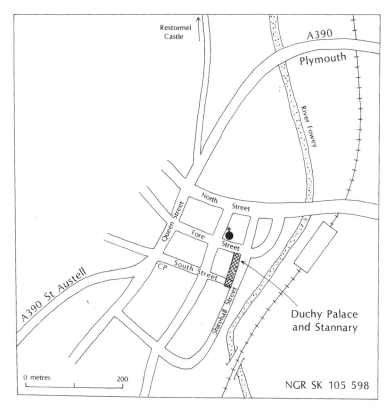

Lostwithiel. Town centre.
Copyright © P.Crew

MORWELHAM, on the east bank of the river Tamar and just in Devon, 3 miles from Gunnislake (NGR SX 445697). A sensitive recreation of a major copper mining port. Museum, bookshop, waymarked walks, underground trips (by train) - plenty to occupy for 2 hours, or all day.

BERE ALSTON MINES, Devon - South Hooe SX 422655, South Tamar SX 435646, Ward Mine SX 426677. Lead and silver mining from 13th to 19th century.

WEIR QUAY SMELTING WORKS, Near Bere Alston, Devon: SX 432650.

Information from Professor N. J. G. Pounds, Dr T. A. P. Greeves and Charles Blick.

91

SITE No. 13

Charterhouse on Mendip

Charterhouse

1 Name of site:

Charterhouse on Mendip, Somerset. Scheduled Ancient Monument.

2 Metal:

Lead.

3 General history of area:

The economic mineral deposits of the Mendip Hills are confined mainly to the central area and flanks of the Blackdown, North Hill and Pen Hill periclines. The main ore is galena, lead sulphide, which occurs in thin veins or lodes. The vicinity of the Blackmoor at Charterhouse has been mined intermittently, probably from pre-Roman times, up to the 19th century. Charterhouse and Ubley rakes are believed to have been worked in Roman and, possibly, medieval periods of mining. Elsewhere in the area, the last era of nineteenth-century industry by its very nature has obscured or destroyed earlier evidence of the extensively documented medieval and post-medieval working.

4 Remains and dating:

The Charterhouse and Ubley rakes consist of several lengths of deep rock-walled gullies on the adjoining hilltop to the south of the Blackmoor Valley. Nineteenth-century working also intruded into this area, leaving the shafts and spoil identified on the Plan of the Rakes, and also the round platform and central stone pivot of a nineteenth-century horse whim presumed to have worked the New Shaft at site 4. The visible surviving remains north of the hill at Blackmoor, and further west at Velvet Bottom and Charterhouse Valley all relate to the 19th century. Of the six separate areas of nineteenth-century dressing floors located on the accompanying plans, two at Blackmoor, one at Ubley and three identified at Charterhouse, the most prominent remains of round buddles are seen just east of the Blagdon-Priddy road, at Charterhouse 3 and settling tanks at Blackmoor 1. The dams and levels of the tailing ponds in Charterhouse Valley are also identifiable, although floodwater sweeping down the valley on the night of 10 July 1968 left some remains covered with silt which, previously, had been easily located and completely removed others. The most recognisable features of the two smelting areas at Charterhouse 1 and Blackmoor 2 are the great lengths of disintegrating, partly restored parallel flues at Blackmoor and huge cubes of abandoned furnace slag overlooking the site of the silted-up reservoir. The reservoir itself is impounded by banks of slag which only very slowly are being overgrown by the vegetation of lead-tolerant species.

5 Location:

In the parish of Priddy near Blagdon, Somerset. Main site NGR ST 488548 to 507561.

6 Accessibility:

Open to the public. Somerset County Council have laid out a guided walk with posts and interpretation boards to the Blackmoor, Velvet Bottom and Charterhouse Valley areas. There is a public footpath to the rakes with a stile giving direct access to the adjoining Blackmoor Valley.

7 Ownership:

The Blackmoor and Charterhouse Valley areas are in the ownership of Somerset County Council while the Charterhouse and Ubley Rakes and Velvet Bottom are in private ownership.

8 Permission required to visit:

None required.

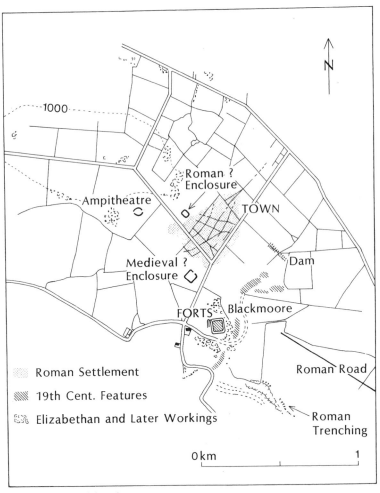

Charterhouse - mining site.
Reproduced by kind permission of Prof. G.D. Barri Jones

9 Sketch plans of site:

Shown on pages 96 and 97.

10 History of working at the site:

It was in the Roman period that Mendip is first definitely known to have been worked for lead with mines at Charterhouse, Priddy and Green Ore, although earlier activity can be inferred from a number of factors. The surrounding region is rich in artefacts and earthwork monuments of the Iron Age period and lead artefacts have been found in a nearby early Iron

Age village. The date of the earliest inscribed ingot of lead in AD 49 (Claudius) is only a few years after the actual invasion of Britain which implies pre-Roman working. Some 19 ingots have been discovered dating up to AD 164-169 (Marcus Aurelius Antonius). They have been found locally on Mendip itself, Bristol, Bath and in Normandy and near Boulogne on the other side of the Channel. Examples can be seen in the museums at Wells and Bristol.

Mendip lead has a high content of silver which the Romans exploited. Little remains of the actual Roman workings apart from the Ubley and Charterhouse Rakes. Traces of a considerable infrastructure can be seen.

Mining is thought to have continued after the Romans left but there is no direct evidence until the 12th century when in 1182 the whole manor of Charterhouse was granted to the Carthusian Priory at Witham, near Frome, by Henry II for use as a grazing area.

The first documented evidence of lead mining is in 1189, with further charters in 1235 and 1283. Production was generally small but by the end of the Middle Ages activity increased and a complex body of laws is recorded showing that the mines were organized into four divisions or 'Liberties' which still can be identified today by their respective smelting sites at Charterhouse, Chewton, East Harptree and St Cuthberts. Ten basic laws were drawn up to cover all aspects of licensing, justice and safety with contraventions being settled by a 12 man jury who met twice a year.

Activity peaked between the years 1600 and 1670. By 1685 coal-fired reverberatory furnaces were being used in Bristol. By the end of the century decline had once again set in due mainly to exhaustion of the easily accessible ore and primitive methods of working. Gunpowder in the 1680s helped with extraction but production was not really improved as the old workings were largely exhausted. The output for 1700-1710 was only one-tenth of that for the previous century. In the early 19th century re-working of the wastes was tackled. There was 25 per cent of lead in some of the slags and 40 per cent of lead in some of the slimes. A brick smelter was erected in 1824 but in the 1840s Cornish mining engineers became interested through their tin mining expertise. Buddle pits and reverberatory furnaces were introduced with the new technology. The slags were washed and crushed and smelted in forced-draught furnaces. The Mendip Hills Mining Company began operations in 1844. Speculative mining was unsuccessful so work concentrated on reworking of old slimes and slags.

Treffry and Co. of St Austell took over the company in 1861 and continued smelting operations, introducing the Pattinson process for the extraction of silver (the remains of the plant were destroyed in the repair of roadworks after the 1968 flood). Smelting was given up in 1878 at Charterhouse but dressed material continued to be sent to Bristol for processing until 1885 when all work ceased due to depletion of re-workable deposits and the low price of lead. Smelting continued at the St Cuthbert's site intermittently until its final closure in 1908.

(Adapted from Somerset County Council Information Leaflet)

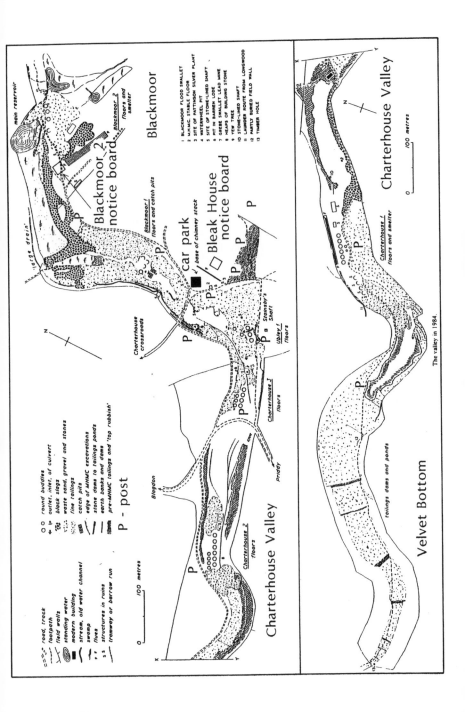

The valley in 1984.

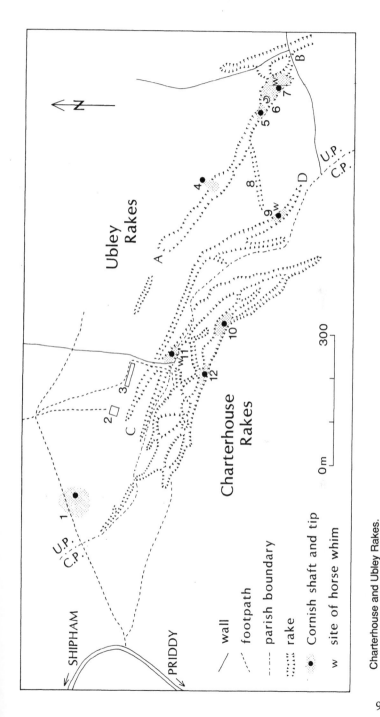

Charterhouse and Ubley Rakes.
These three diagrams are reproduced by kind permission of Dr W. I. Stanton and A. C. Clarke from their paper " Cornish Miners at Charterhouse-on-Mendip", in *Proceedings of University of Bristol Spelaeological Society*, 1984, Vol. 17 (1), pp. 29-54.

11 References:

Buchanan, A. and Cossons, N. (1969) Industrial Archaeology of the Bristol Region, David & Charles, Newton Abbot.

Elkington, D. (1976) 'The Mendip lead industry', in Branigan, K. & Fowler, P.J., The Roman West Country, David & Charles, Newton Abbot.

Gough, J.W. (1967) 'The Mines of Mendip', David & Charles, Newton Abbot.

Hawtin, F. (1970) 'Industrial archaeology at Charterhouse on Mendip', Industrial Archaeology, 7(2), 171-175.

Kingsbury, A.W.G. (1941) 'Mineral Localities on the Mendip Hills, Somerset', Mineral Mag., 26, 67-80.

Stanton, W.I. & Clarke, A.G. (1984) 'Cornish miners at Charterhouse-on-Mendip', Proc. Univ. Bristol Spelaeol. Soc., 17(1), 29-54.

Contact Addresses:
For General information about the archaeology contact:

The County Archaeologist, Planning Department, County Hall, Taunton, TA1 4DY. Tel: 0823 255426.

Or, for other information, contact:

The Mendip Hills Warden, Charterhouse Outdoor Activity Centre, Blagdon, Bristol. Tel: Blagdon 62338.

12 Adjacent sites of interest:

ST CUTHBERT'S LEAD WORKS, Priddy, Avon: ST 545505. Fragments of a condenser, chimney bases and extensive foundations and flues from final effort of early 20th century in reworking earlier lead slags.

EAST HARPTREE LEAD WORKS, Smitham Hill, Avon: ST 556547. Surface features largely obscured by encroaching forestry plus the only surviving Mendip smelting chimney.

CHEWTON MINERY, near Priddy, Avon: ST 547515. The main feature of this former smelting site is the roadside reservoir although leats, buddles and a flue can be seen but earlier remains have been obscured by forestry development.

SHELDON BUSH AND PATENT SHOT TOWER, Cheese Lane, Bristol, Avon: ST 595729. Twentieth-century lead shot tower replacing eighteenth-century tower built by William Watts who developed and patented the process.

Information from Mrs J. Day, 3 Oakfield Road, Keynsham, Bristol BS18 1JQ, Professor G. D. Barri Jones, and Somerset County Council, County Planning Officer to whom we are indebted for permission to publish this information.

SITE No. 14
Leadhills, Lanarkshire and Wanlockhead, Dumfriesshire

1 Name of site:

Leadhills and Wanlockhead.

2 Metal:

Gold then lead.

3 General history of area:

The gold mining area in southern Scotland stretched from Linlithgow in the north to the Leadhills and Wanlockhead in the south, with a width of about 10 miles. All the gold came from the surface, mainly near streams. Medieval miners at Leadhills do not seem to have found gold below the surface, but bronze picks are recorded from surface workings which suggest that goldworking may have been quite early in this area.

Probably the best known of the publications on the history of this area is by the Rev. J. Renoir Porteous with the title 'God's Treasure House in Scotland', dated 1876. Finds of gold ornaments 'Celtic sixth to ninth centuries' are noted, some now in the National Museum of Scotland in Edinburgh. There is a document of 1239 concerning a lawsuit which refers to a lead mine at 'Glengower', on Crawford Muir (an area around Glengonnar Water, running north from Leadhills into the river Clyde). References appear in Royal Accounts of 1488 to 1513 to goldworkings at Crawford Muir, one nugget weighing 2lbs 3oz being found in 1502. By the middle of the 16th century gold was being produced in Leadhills, and the neighbouring area, sufficient for Queen Mary's gold coinage and possibly her crown. Later that century the surface deposits of natural gold were being exhausted and the working of copper and lead took over. These also contained precious metals which were obtained from such ores by cupellation (Patrick, 1878).

Lead mining on the WANLOCKSIDE developed to a commercial scale in the 17th century, and lasted until the 1930s. The London Lead Company started work in 1710 and in 1721 the Friendly or Quaker Company joined in. A single enterprise was formed in 1755 and lasted until 1842 when the Duke of Buccleugh took control, working the mines until 1906 when the Wanlockhead Lead Mining Company took over. After World War I business did not prosper and the mine closed in 1934. An attempt to reopen the mine in the 1950s was unsuccessful and operations ceased in 1959.

4 Remains and dating:

There are no visible structural remains but there are two areas of medieval metal extraction: RESPIN CLEUGH NGR NS 901158 - alluvial gold from stream; particles are still found there, and LADY MANNER'S SCAR NGR NS 881158 - visible from the road: early slag found there. Both are evocative of the wild nature of the Scottish Southern Uplands wherein the four gold bearing rivers, or Waters, are Wanlock, Mennock, Shortcleugh and Gonar (now Glengonar).

5 Location:

LEADHILLS NGR NS 876153 at junction of B797 and B7040 from Elvanfoot. WANLOCKHEAD NGR NS 864128 on B797 road to Mennock and Sanquhar.

6 Accessibility:

It is ESSENTIAL for the maximum enjoyment of, and information on, the area to call first at the Museum of Scottish Lead Mining, Wanlockhead Museum Trust, Goldscaur Row - on the site of the early goldworkings - Wanlockhead. It is open daily during the summer months (14.00-16.00) during June, and 11.00-19.00 during July and August. Here is the Trust's collection of mining and social relics, with a display of many minerals and a range of publications on local mining and social history.

7 Ownership:

Visitors are asked to keep to the roads and main tracks and to respect private property. Most sites can be seen therefrom. Mine heaps and shafts to be approached, if at all, with care.

8 Permission required to visit:

The museum provides a guide book with routes to follow, maps and notes on individual sites.

9 Sketch plan of site:

The museum's book provides all details necessary. Maps of local area are shown on page 102 and 103.

10 History of working at the site:

Incorporated in section 3.

11 References:

Renoir Porteous, The Rev. J. (1876), God's Treasure House in Scotland.
Mitchell, J. (1919) The Wanlockhead Mines.
Smout, T.C. (1967) Lead Mining in Scotland; 1650-1850.
Patrick, R.W.C. (1878) Early Records Relating to Mining in Scotland, Edinburgh.

12 Adjacent sites of interest:

None near enough: the area of the two villages provides a good day's outing.

Information from Dr E.A. Slater, Dr P. Swinbank, R. Ellam, W.S. Harvey and Charles Blick.

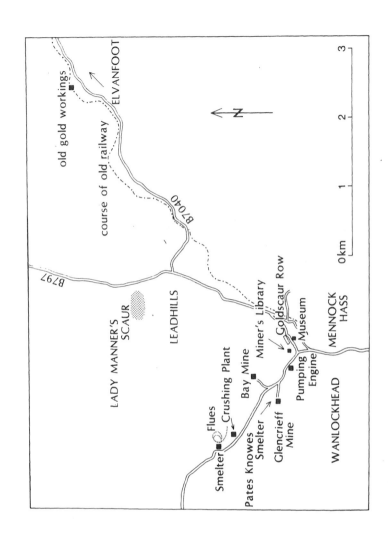

Wanlockhead and Leadhills.
Courtesy of the Museum of Scottish Lead Mining, Wanlockhead Museum Trust

Leadhills and Wanlockhead. Mining sites.
Courtesy of Museum of Scottish Lead Mining,
Wanlockhead Museum Trust

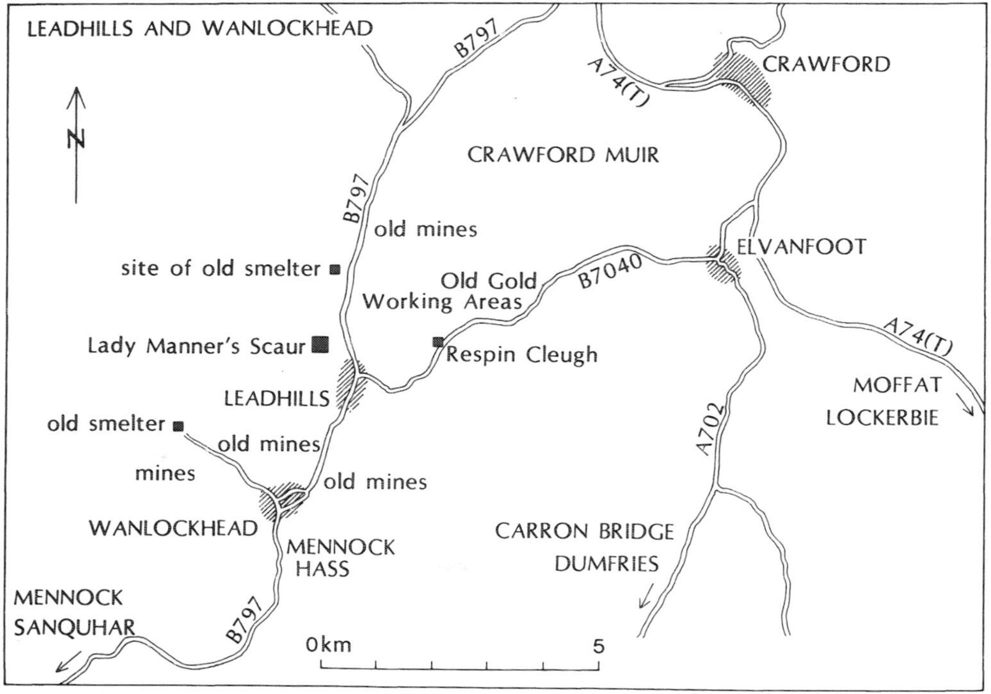

LEADHILLS AND WANLOCKHEAD

CRAWFORD

A74(T)

ELVANFOOT

MOFFAT
LOCKERBIE

A702

B7040

CRAWFORD MUIR

Old Gold
Working Areas

Respin Cleugh

old mines

B797

B797

site of old smelter

Lady Manner's Scaur

LEADHILLS

old mines

old mines

CARRON BRIDGE
DUMFRIES

old smelter

mines

WANLOCKHEAD

MENNOCK
HASS

MENNOCK
SANQUHAR

B797

N

0km   5

SITE No. 15
Great Orme

Great Orme

1 Name of site:

Great Orme, Llandudno, Gwynedd.

2 Metal:

Copper.

3 General history of area:

The Great Orme is a headland of Carboniferous Limestone that rises to a height of 679ft situated to the west of the resort of Llandudno. Last century saw the rapid growth and demise of an important copper mining industry here. A number of mineral veins have been identified; they predominantly strike north-south with steep dips to the west and are most developed in the Pyllau valley to Bryniau Poethion areas. At least four major lodes together with several minor parallel and occasionally oblique trending veins have been exploited. The limestones in the vicinity of these veins have become strongly dolomitized, providing host areas for primary sulphide mineralization which is controlled both stratigraphically and structurally. In fact particular areas could be more suitably described as a stockwork. The principal ore chalcopyrite exists at surface and continues up to depths of 200m, alteration and secondary (supergene) enrichment has given rise in the near surface zone and in localized horizons to carbonate ores, mainly malachite and to a lesser extent azurite. Tenorite, native copper and small but profitable quantities of galena were also recorded. A characteristic and important feature of the dolomite is its tendency through weathering to produce a rotted granular deposit, which must have made early extraction of copper ores a simple exercise.

4 Remains and dating:

Construction work for the reclamation scheme uncovered exposures of deeply dissected vein rock around the Vivians shafthead; this disclosed what now is accepted as an early mined land surface. Worked out spoil-filled veins with rotted dolomite wall rock are clearly discernible. Excavations by Gwynedd Archaeological Trust have produced numerous bone and stone implements with associated tool marks in the rotted dolomite and grey shale horizons. Reddening of dolomite wall rock together with locally abundant residual charcoal suggests fire-setting was being employed at surface. Charcoal obtained from a sealed context within a confined gallery on the eastern face of the exposure has provided the earliest date of 3370 +/- 80 (CAR-1184) calibrated 1747-1535 BC. Surface excavations are continuing.

5 Location:

West of the town of Llandudno, Gwynedd. NGR SH 780 820.

6 Accessibility:

The area of archaeological interest is situated in a fenced enclosure with locked access. There is no public admittance, however, reasonable views can be obtained from the high ground surrounding the site.

7 Ownership:

Aberconwy Borough Council. A scheme is presently at hand to develop the site as an interpretive facility, with limited opening. The lessee will be Great Orme Mines Ltd.

8 Permission required to visit:

Permission to visit the site can be arranged by the Great Orme Exploration Society. This will be for a limited period until the site is leased by the developer. Enquiries Mr C.A. Lewis, Gwynedd, Tyn-y-coed Road, Great Orme, Llandudno, Gwynedd, LL30 2JY. Tel No. (0492) 79202.

9 Sketch plan of the site:

Shown on page 106.

Great Orme's head.

Courtesy of
Dr W.D.F. Smith

KEY

1 Roman shaft
2 Treweek's shaft
3 Higher shaft
4 Owen's shaft
5 Engine house
6 Pyllau shaft
7 Pyllau farm
8 Vivian's shaft
9 Washing floors
10 Tyn y Fron shaft
11 Ty Gwyn shaft
12 Kendrick's cave

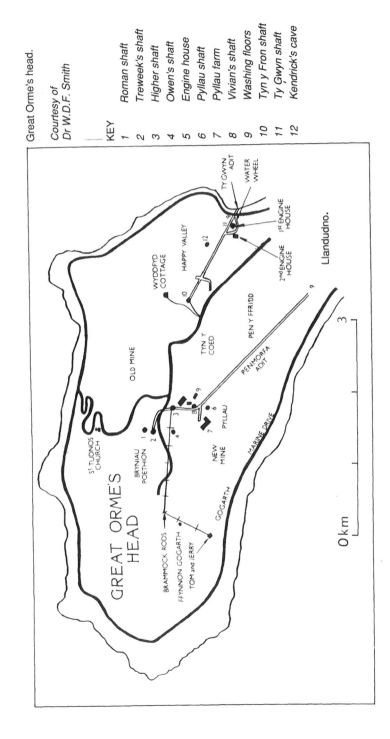

GREAT ORME'S
HEAD

St TUDNOS
CHURCH

BRYNIAU
POETHION

BRAMMOCK RODS

FFYNNON GOGARTH

TOM and JERRY

GOGARTH

NEW
MINE

MARINE DRIVE

OLD MINE

PYLLAU

PENMORFA
ADIT

TYN Y
COED

PEN Y FFRIDD

WYDDFYD
COTTAGE

HAPPY VALLEY

TY GWYN
ADIT

WATER
WHEEL

1st ENGINE
HOUSE

2nd ENGINE
HOUSE

Llandudno.

0 km   3

10 History of working at site:

The earliest documentary evidence dates from 1692 when a lease was taken out to mine copper in the Pyllau valley; even at this time, former mining was known. Operations progressed intermittently during the 1700s, with ore production not really taking off until the 1820s which saw a period of increased activity up to the late 1850s. By the 1870s all of the mines were in rapid decline, pumping operations curtailing to allow flooding of the deep workings that had attained depths of 70m below sea level.

In the years 1831 and 1849 two important discoveries shed the first light for the existence of early mining on the Great Orme. Many contemporary accounts of the time described how prospecting miners broke into workings described as Roman or belonging to the Old Welsh or Celtic miners (Williams, 1979; Smith, 1989). Spectacular stalactite formations, stone hammers weighing between 2-40lb, implements of bone from ox, swine and deer, and a few items of bronze and charcoal remains from fires were reported. Other instances of stone hammers or mauls being found amongst mine waste filling the old workings are not uncommon.

During 1938-39 the eminent archaeologist Oliver Davies visited the site to investigate the evidence earlier reported; he records further stone hammers from the Pyllau area and also from a limited excavation of a settlement site at the West Shore near to the present Gogarth Abbey. No underground examinations could be made as all entrances to the mines had been effectively sealed soon after their closure. With this albeit limited information he ascribes a Roman date on the basis of pottery and coins found in the mining area (Davies, 1948). This view was regarded until comparatively recently as the age of the earliest mining here.

Duncan James's pioneering researches during the late 1970s under Bryniau Poethion really brought the field study of early mining to a front, inspiring investigation of at least some of the many now radiocarbon dated Bronze Age mining sites in this country. His excavations revealed a system of generally horizontal galleries up to 50m in length at vertical depths to about 30m, and containing unmodified cobble hammerstones, bone implements and calcite flowstone (speleothems), clearly capping stratified sequences of characteristic mine spoil incorporating fragments of charcoal, suggestive of firesetting. This material provided a radiocarbon date of 2940 +/- 80 (HAR-4854) calibrated 1300-1020 BC, implying middle Bronze Age mining operation.

During 1987 a mining survey of all workings 20m below surface as part of a reclamation scheme at the site provided further prehistoric evidence. A complex system of interconnecting multiple veins and laterally extending ore bodies were encountered. In particular locations exhibited features and artefacts similar to those previously commented upon. When considered collectively and backed by radiocarbon dates, four of which now exist, such criteria can be used to provenance these sites to the Early-Middle Bronze Age.

11 References:

Davies, O. (1948) 'The copper mines on the Great Orme's Head, Caernarvonshire', Archaeologia Cambrensis, 100, 61-66.

James, D. (1988) 'Prehistoric copper mining on the Great Orme Head, Llandudno,Gwynedd', in Aspects of ancient mining and metallurgy, J. Ellis Jones, (ed.), UCNW, Bangor, pp 115-121.

Jenkins, D.A. and Lewis, C.A. (1990) 'Prehistoric mining for copper in the Great Orme, Llandudno', British Archaeology Reports.

Smith, W.D.F. (1989) The Great Orme Copper Mines, Creddyn Publications, Llandudno.

Williams, C.J. (1979) 'The Llandudno Copper Mines', British Mining No. 9, Monograph of the Northern Mines Research Society.

12 Adjacent sites of interest:

PARYS MOUNTAIN COPPER MINE, Anlwch, Anglesey: spectacular remains of the largest copper working in the world in the 19th century.

SYGUN. Nineteenth-century Copper Mine, Beddgelert, Gwynedd, Museum.

Information from C. A. Lewis, Great Orme Exploration Society, 'Gwynedd',Tyn-y-coed Road, Great Orme, Llandudno, Gwynedd. We are indebted to Dr W. D. F. Smith for information and permission to publish the site plan from his book 'The Great Orme Copper Mines' ISBN 0951 4051 01.